## What People

"Brian and Kristine have created powerful connections between the heart and mind and the learning of math. It is interesting to consider that the struggle and fear of math may have less to do with numbers themselves."
—Laura Henderson, Founder and CEO of Epiphany Learning

"*I'm Just Not a Math Person!* is an insightful and deep read. My mind was blown countless times as I read, making connection after connection to my own life experiences both as a student and as a teacher. The "aha" moments broadened my understanding and helped me gain valuable insight into the minds of my students, and the real-life classroom examples helped to solidify my understanding and connect the ideas to my classroom. The knowledge and understanding I have gained will help me be more successful as I move forward and help more students access their mathematical potential."
—Darral Sessions, Math Curriculum Specialist at Wichita Public Schools

"Peters and Hobaugh present a great case for teaching to the whole child, especially within the context of math – an academic subject unfortunately feared by so many. The concepts and applications, presented in a conversational and approachable manner, make it an enjoyable read for any teacher of K-12 mathematics."
—Dr. Catherine Oleksiw, Managing Principal at Measured Transitions LLC

"Brian and Kris have done a fantastic job of making math anxiety and fear understandable. They unpack the sociological elements of fear that intersect with how we all learned math, highlighting the WHY, and then provide usable and understandable solutions that always tie back to the research. This will benefit any teacher or parent by providing insight into how the brain learns and deals with "math fear." This book is a must read for teachers of mathematics at all levels."
—Chris Castillero, Mathematics Specialist at the Rhode Island Department of Education

"I loved math as a kid; however, I had friends who would literally get sick at the sight of a math problem. Brian and Kristine's book shows that math is not a bad word and gives the truth about how to navigate the fear of math."
—Jody Harris, Award-Winning Inventor, Best-Selling Author, and Motivational Speaker

"The authors masterfully discuss the heart of mathematics classrooms in our country: the constraints and shortcomings of "The System," the realities of our students and society, and our duty and commitment as leaders and educators. With authentic classroom excerpts and lesson ideas, the reader is left inspired with relevant applications built on the foundations of NCTM's Process Standards and the Eight Standards for Mathematical Practices."
—Julia Baucum, Mathematics Content Area Specialist, Scottsbluff Public Schools

"A great book that is a must read for all math teachers, new or experienced. Brian and Kris provide excellent ideas to get students engaged in, and enjoying, math. I was inspired by the book to make changes in my classroom and implement strategies to reach out to all of my students. While we cannot make everyone love math, we can attempt to take the fear of math out of our classrooms when working with our students."
—Vicky Wood, Math Coordinator for the Glenwood Community School District

*"Nicely done! The concept is definitely interesting and worth considering."*
—Stefanie Rothstein, Academic Coach at Rocori Area Schools

*"Brian and Kris have written a compelling text about WHY math instruction needs to change in our schools. Anecdotal stories provide a picture of HOW to shift practice. Brian and Kris's commitment to making math fun, interesting, and meaningful will inspire readers to join in this mission."*
—Lori Loehr, Instructional Coach at the School District of Menomonee Falls, WI

*"Finally a book that addresses why math is scary for so many people! Peters and Hobaugh do a wonderful job of taking a complex problem and breaking it down into manageable pieces that can be understood by school officials, parents, and even our students. With a shared perspective, hopefully we can each do our part to help our young students rethink what math instruction should look and feel like, as well as reduce fears that have long been associated with learning math."*
—Clarissa Burt, Entrepreneur & International Media Personality

*"We live in an age when math is the language of success in many fields. Missing the core language is tantamount to being marginalized in life and put outside the wider range of possibilities that pay better. This book deals with human issues – it's a must read!"*
—LeiLani Cauthen, CEO Learning Council & Knowstory;
Author, *The Consumerization of Learning*

*"'I'm Just Not a Math Person!' delves into the reasons why so many students fear math! In addition, the authors provide numerous reasons why the education system continues to perpetuate this cycle of fear. The good news: the book provides concrete examples of how to shift learners from fear into problem-solving and accelerating the learner's math skills. The "light bulb" moments provide practical takeaways for teachers to immediately implement."*
—Dr. Heidi Hahn, Director of the Paul Bunyan Education Cooperative (Brainerd, MN)

*"In a time where anxiety is high for teachers, students, and the community, this book will empower all to conquer their fears, knowing that they are not alone! 'I'm Just Not a Math Person!' addresses students' and teachers' fears in ways that make you feel their potential and possibilities."*
—Dr. Angela G. Ford, Mathematics Curriculum Specialist at Milwaukee Public Schools

*"I loved the idea of looking at fear and how that leads to students being apprehensive or downright negative towards math. There are great thoughts on math fear and how teachers can get students to let go of their fears. Plenty of practical advice to make classrooms a place where students can feel safe and teachers can engage in best practices in mathematics instruction."*
—Nate Rosin, K-12 Mathematics Coach at the Sun Prairie Area School District, WI

# I'm Just Not a Math Person!

## Recognizing, Understanding, and Managing the Fear of Mathematics

Other books by Brian A. Peters

*The METUS Principle:*
*Recognizing, Understanding, and Overcoming Fear*

*The Pocket Guide to Leadership:*
*The 9 Essential Characteristics for Building*
*High-Performing Organizations*

# I'm Just not a Math Person!

## Recognizing, Understanding, and Managing the Fear of Mathematics

**Brian A. Peters, MEd, MSA, MBA**
**Kristine E. Hobaugh, MEd**

Henschel HAUS publishing, inc.
MILWAUKEE, WISCONSIN

Copyright © 2019 by Brian A. Peters

All rights reserved. No part of this publication may be reproduced, distributed, or transmitted in any form or by any means, including photocopying, recording, or other electronic or mechanical methods, without the prior written permission of the publisher, except in the case of brief quotations embodied in critical reviews and certain other noncommercial uses permitted by copyright law. For permission requests, write to the publisher, addressed "Attention: Permissions Coordinator," at the address below.

Published by
HenschelHAUS Publishing, Inc.
2625 S. Greeley St. Suite 201
Milwaukee, Wisconsin 53207
www.henschelHAUSbooks.com

HenschelHAUS books may be purchased for educational, business, promotional, or professional use or for special sales.
Quantity discounts are available through the publisher.

ISBN: 978159598-570-5 (Paperback)
E-ISBN: 978159598-683-2
LCCN: 2017914452
Cataloging in Publication information pending

Cover design and illustrations by Melisa Cash

# Table of Contents

Acknowledgments .................................................................. i

Introduction: Brian A. Peters .............................................. 1
Introduction: Kristine E. Hobaugh ...................................... 3

Preface ................................................................................ 9

**Fear, Development, and Mathematics** .......................... 11
    Fear ............................................................................... 11
        How Fear and Anxiety Impact Learning ................ 13
        How Fear Works ....................................................... 17
        Fear's Effect on Thinking ........................................ 18
        Fear and Anxiety and Their Effects on Development ..... 19
    Development ............................................................... 26
        The Other End of the Spectrum ............................ 26
        Developing a Learner's Mindset ........................... 29
        Teaching vs. Learning ............................................ 31
        Motivation, Goal-Setting, and Growth ................. 36
    Mathematics ................................................................ 45
        The Practitioner's Lens .......................................... 45
        Why Do People Hate Math? .................................. 46
        Recognize, Understand, Manage:
            The Value of Precision ..................................... 51
        Kris's Classroom ..................................................... 51

**The Foundation for Meaningful Change
in the Learning Environment** ...................................... 61
        Recognize, Understand, Manage:
            A Positive, Student-Centered Approach ........ 62
        What are the First Steps in Creating a
            Student-Centered Classroom? ......................... 65
        Kris's Classroom ..................................................... 67

**The System** ............................................................................. 73
    Recognize, Understand, Manage ................................................. 73
    Kris's Classroom ............................................................................ 75

**How Do We Develop Math Learners?** ................................... 87
    Recognize, Understand, Manage ................................................. 87
    Student-Centered vs Teacher-Centered Teaching ................. 94
    Kris's Classroom ............................................................................ 95
    Productive Struggle .................................................................... 103
    CAN vs CAN'T ............................................................................. 109
    Closing Exercise: The Ideal Classroom ................................... 112
    Kris's Classroom .......................................................................... 113

**Afterword** .............................................................................. 117

**References and Resources** ................................................... 119

Addendum A: Practitioner's Summary ............................................... 123
Addendum A: Eight Standards for Mathematical Practices ....... 125
Addendum B: Business Card Activity ................................................. 135

About the Authors ................................................................................. 139

# ACKNOWLEDGMENTS

**BRIAN:**

*"To see a change, you need to become a living, breathing asset to everyone you know and a true advocate for everything that you believe in."*
—Jody Harris, Award-winning Inventor, Best-selling Author, and Motivational Speaker

So many things in our lives come and go, but the experiences we share with others endure. When we have an opportunity to make a positive difference in the someone else's life, we should embrace that opportunity. I cannot think of a profession or field that embodies this perspective better than education, and it is for this reason that I love being a part of it.

Shortly after the release of my award-winning book, *The METUS Principle*, my friend Christine Lund asked me if I thought the METUS framework could be used to explain why so many kids are afraid of math. At that moment, the concept for *I'm Just Not a Math Person!* was born.

After several months of working on what would eventually become parts of this book, I realized that the challenges we face in math are not just teacher or administrator problems, they are not just parent problems, and they are not just student problems. As a result of this realization, I knew that this book needed to take a very complex topic—math achievement—and simplify it in a way that was helpful to not only teachers and administrators, but to parents and students as well. As I considered different perspectives, I knew that I needed a strong practitioner's lens. I needed someone who could take a deeper dive, provide concrete or working examples, and offer actionable guidance.

Working in and around education for many years, I have been truly blessed to meet some wonderful people and gifted minds. This statement couldn't be any more true of Kris Hobaugh. Kris has both the desire and the unique ability to make those around her better, which is why I knew she would be the perfect person to collaborate with on this project. While it took a bit of convincing, she finally said yes—and I am truly thankful that she did.

On behalf of both Kris and myself, I want to thank our amazing readers and those who have contributed in a meaningful way to the success of this endeavor:

# Acknowledgments

Julia Baucum
Dr. Steve Bialek
Stephanie Blue
Dr. Melissa Boston
Sara Brown
Clarissa Burt
Melissa Cash
Chris Castillero
LeiLani Cauthen
Amy Cowell
Rege D'Angelo
Lesley DeMartini
Dr. Kelly Edenfield
Dr. Angela Ford
Dr. Melissa Freiburg
Dr. Tonja Gallagher
Jenny Gomez
Dr. Stephanie Gottwald
Dr. Heidi Hahn
Jim Harrington
Jody Harris
Laura Henderson
Kira Henschel
Holly Ingram
Lori Loehr
Christine Lund
Susan Niette
Dr. Catherine Oleksiw
April Pforts
Nate Rosin
Stefanie Rothstein
Darral Sessions
Dr. Cathy Seeley
Jill Swissa
Liz Tierney-Olson
Mark Tipton
Daphne Williams
Vicky Wood

## Models and Images

Crystal Ayad, Jailine Barbosa, Stephanie Blue, Rachelle Cruz, Briannah Erbs, Reagan Hoffman, Richard "Hutch" Hutchinson, Chaniece Jeffery, Raysha Marrero, Amara Marshell, Friedericke "Fred" Marshell, Miana Marshell, Kensington Peters, McKinley Peters, Missy Peters, Saradhi Saripalli, Blake Schultz, Rabecca Siang, Hayden Ton, and Taylor Wharton.

Lastly, I want to thank my amazing wife, Missy, and our daughters, Kensington, McKinley, and Leilani. I have watched my wife work tirelessly as an elementary school teacher to bring out the best in her students and to help them understand their potential. It is my hope that all kids, including our girls, will have teachers in their lives who want what is best for them from the very bottom of their hearts.

# ACKNOWLEDGMENTS

**KRIS:**

I want to start by saying thank you to our readers. Whether you are an administrator, teacher, parent, student, or someone with math anxiety, we hope that you can enjoy and relate to our experiences in this book. Math is language—active and fun if you are teaching and learning it the correct way.

When Brian came to me with this idea about writing a book to help people tackle their math fears, I was hesitant because while I love math, I am not a writer. I even told Brian that I love to speak, but please don't ask me to write. So, Brian, thanks for pushing me out of my comfort zone and encouraging me to write down my experiences to help others.

I also want to thank my son, Nathanial, for allowing me to experiment on him and share our family journey of mathematics in our home.

A special thanks to my mathematics education professor, Nina Girard, who helped me shape my knowledge of teaching and learning in college.

Thank you as well to all the friends, family, teachers, and students who have been a part of my math journey over time.

I also want to thank my colleagues who have taught me so much over the years: Sandy Bartle Finocchi, David Dengler, Dr. Bill Hadley, Amy Lewis, and Cassie Reynolds, and Dr. Steve Ritter, to name just a few. Sandy, thanks for taking a chance on me and reminding me to put on my "big girl pants" to get the work done.

Lastly, I need to thank my husband, Don, who over the past year has sacrificed countless late nights, weekends, and even most holidays so that I could spend time writing and rewriting this book. As a mason, carpenter, and teacher, he always sees the value of mathematics and has helped me many times to make real-world connections.

I am truly blessed to have all the above-mentioned people in my life to support me professionally and personally. I couldn't have completed this without each of you!

This book is about reducing fear and anxiety far too often associated with learning math.

We will discuss the importance of building authentic relationships, providing personalized support, and helping students and teachers reach their goals.

The less fear and anxiety students or teachers hold onto and the more they feel supported and valued, the harder they will work and the better they will perform.

**THE PROBLEM:**
*Too many students dislike, hate, and/or fear mathematics.*

**THE RESULT:**
*Far too many students struggle and underachieve as math learners.*

**THE IMPACT:**
*There are life-long consequences, both professionally and personally.*

# The Truth:

*You are capable of learning math!*

*Your students are capable of learning math!*

*Your sons and daughters are capable of learning math!*

*We are ALL "math people"*
*under the right learning conditions.*

CONTENT HAS BEEN COLOR-CODED FOR FOCUS
ON THE FOLLOWING FOUR AUDIENCES:

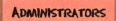

**ADMINISTRATORS**

**TEACHERS**

**PARENTS**

**STUDENTS**

IN ADDITION, YOU'LL FIND COLOR-CODED BOXES FOR:

**GENERAL INFORMATION**

AND

**LIGHT BULB MOMENTS**

# Introduction

### Brian A. Peters

Have you ever heard someone say, "I am just not a math person..."?

What about, "I am a right-brained person and not a left-brained person, which is why I am not good at math..."?

And then there's, "My mom and dad weren't math people either."

Perhaps you have even said those words yourself!

Have you ever heard parents try to justify their child's struggle in math? "I wasn't a math person either." Or, "Honestly, I didn't really like math either. I'm pretty successful. You don't have to be good at math to be successful."

Even administrators and teachers fall into the trap. "Our district has always had really low math scores." Or, "Most of my students were years behind before they even got to my classroom. I'm just trying to help them pass."

As you read these statements, I can just about guarantee that you were thinking of someone, or perhaps several someones, you have heard utter those very words. Why do I say this? Because at any given time, there are MILLIONS of American students who hate, despise, loathe, or fear mathematics.

> "Be kind. Everyone you meet is carrying a heavy burden."
> Ian MacClaren, *Beside the Bonnie Brier Bush*

As you get ready to dive into our book, I want to challenge you to consider the information you read from multiple perspectives—the four hats of administrator, teacher, parent, and student.

The struggles students face in math are real. There are many reasons that students experience difficulties, and in many cases a lack of success, in math. Whether you are an administrator, a teacher, a parent, or a student, this book was written to help you better understand how fear and anxiety impact math achievement.

Individually, we are well aware of our own struggles. However, we often fail to realize or appreciate the struggles of those around us. Administrators struggle. Teachers struggle. Parents struggle. Students struggle. To admit struggle is not a sign of weakness. In fact, allowing yourself to be vulnerable and admit struggle is a sign of strength. Strength is also found in stepping up, demonstrating compassion, and having the desire to help others around us.

In my award-winning book, *The METUS Principle*, I offer readers insights into the decision-making process and what I believe to be the most basic catalyst of behavior—fear. *The METUS Principle* was written to help readers recognize, understand, and manage fear in their lives. In *I'm*

# INTRODUCTION

*Just Not a Math Person!*, my colleague, Kris E. Hobaugh), and I will use the METUS framework to describe why students struggle, avoid, and in many cases, hate math. We will also provide examples and strategies that can be applied to help reduce fear and anxiety and foster a learner's mindset.

Oh, and by the way, for the first 20 years of my life, I was someone who claimed I wasn't born with the math gene too. Yes, I hated math.

# INTRODUCTION

### KRISTINE E. HOBAUGH

I decided in high school to become a math teacher because I loved making sense of information and helping others understand and make sense of information around them. I wanted to be a teacher because I enjoyed helping others learn, and I thought I wanted to be an elementary teacher—until Geometry class.

That year I was doubling up in math, taking Algebra II and Geometry in the same year. These were two very different classrooms.

My Algebra II teacher, Mr. Franks, was in his last year of teaching. We sat in rows and columns, taking notes from bell to bell. Homework every night was numbers 1-50, evens or odds. He would entertain questions, but they had to be related to the procedure we learned the day before or the new one we were learning.

I liked algebraic manipulation, and because I was successful I really liked his class. What I didn't learn from Mr. Franks, I learned by reading the textbook and making sense of the worked examples. I noticed quickly that I could also daydream or write notes to my friends during class, because the examples that Mr. Franks wrote on the board were also in my textbook.

Mrs. Patterson's Geometry class was very different. She expected me to participate. She would toss out a task to the class and ask us to write down how we thought it should be solved. We would then share with a partner what we had thus far and then as a full group, try and solve whatever was still remaining. She allowed us to work alone if we chose to, rather than with a partner. She did, however, expect that we would share our work with the whole class if called upon.

Mrs. Patterson never embarrassed or intimidated us, but she asked us to talk about our thinking so that everyone could decide if it made sense. I learned quickly that her tasks could not be found in the book. I would need to pay attention, because I wouldn't have the option to teach myself when I got home. The discussion about math concepts was the focus of her class, not the mindless memorization and practicing of procedures. We took a lot of notes, but we also engaged in the class. The notes we took personalized our learning.

# INTRODUCTION

Even if they liked Geometry, most people probably wouldn't say that proofs were their favorite. I, however, loved proofs. You already knew the first and last statements, so it was the story in the middle that was the puzzle to solve. After each step, Mrs. Patterson asked, "Does that make sense? Why? Explain." If you could convince her and the class that you were correct, she would allow that answer. I clearly remember one day after working with our partners, she asked us to come back together as a group because David had a question. David was the class clown, cute, and popular. I was an underclassman in this class, so I rarely spoke up unless called upon.

David went up to the board to walk through his proof and show why two given lines were parallel. His question was, "Is this the only proof that was correct?" because his partner had solved it very differently.

I did not solve this problem the same way David had, but I was not going to raise my hand and tell him he wasn't correct. Mrs. Patterson noted that my partner and I had solved it differently from David, and she asked us to share our method. My partner, Kristen, went to the board.

As Kristen showed our method, David and the rest of the class listened intently. She finished by saying, "David, I think your method makes sense, but our method also works." David was not happy because he thought there should only be one way to solve problems; that was how it had always been in our previous math classes. There might be various methods of getting there, but there was only one right answer. Mrs. Patterson reminded the class that sometimes we would have multiple methods and sometimes multiple answers that would make sense for problems we would work on in her class.

In Mr. Frank's Algebra II class, David was right—there was only one right answer or method. In Mrs. Patterson's classroom, the focus was on making sense of the mathematics. The other difference between the two classrooms was that one fostered a positive, student-centered approach, while the other focused on covering the content that was in the book. Mrs. Patterson's classroom was a collaborative environment where students would take risks. She encouraged us to develop our own methods and make sense of the tasks in front of us. I only learned to memorize and practice procedures in Mr. Frank's classroom.

Mrs. Patterson made connections to the real world so that we, as students, had context as to why we needed to learn the mathematics. She didn't expect us to just memorize but to be able to write and explain our reasoning. While we also did application problems in Mr. Frank's class, they were from the textbook. Mrs. Patterson had us go home and measure a room in our house to find the area, surface area, and volume of the room. She did her best to connect to our interests as well.

Those two teachers in my life had two diametrically opposed approaches. As much as Mrs. Patterson and

## INTRODUCTION

> *Real math problem-solving should be open-ended.*

Mr. Frank helped me learn math, they taught me so much more about different styles of learning. I liked both teachers and both classes (yes, I was a geek)—but I learned to love math in Mrs. Patterson's classroom.

I became a math teacher because of my love for geometry. I earned a Masters in Education with a concentration in Mathematics and Computer Science, and I taught high school mathematics for 10 years in a suburb of Pittsburgh. I facilitated several university math and professional development courses during that time, after which I decided to branch out into educational business. I joined Carnegie Learning, where I am currently the Director of Professional Development, North U.S. During my time at Carnegie, I also returned to school to earn my K-12 administrator's certification.

Throughout this book, my experiences as a teacher, professional, and parent will be shared in sections called "Kris's Classroom."

> *After 21 years of teaching and delivering professional development for mathematics education, I have learned that Mrs. Patterson's approach was student-centered, while Mr. Frank's classroom was teacher-centered.*

# PREFACE

Kris and I had two primary goals in writing this book:

1. To provide an explanation and a window into the minds of those who experience fear related to mathematics, and
2. To provide recommendations and strategies for administrators, teachers, parents, and students to reduce fear associated with math and offer a different narrative for math conversations in our society.

The METUS Principle relies on cross-disciplinary research in an effort to explain motivation by looking at behavior and the decision-making process. For the purpose of this book, it is important to understand the basics of the METUS Principle, which are as follows:

- The word *metus* is Latin for fear.
- Fear is an innate response that accounts for how people think, act, and respond to environmental stimuli.
- Response to fear is dependent on needs and is consistent with Abraham Maslow's *Hierarchy of Needs,* which includes physical safety and security, love and belonging, status and accomplishment, and self-actualization.

- When individuals experience situations that trigger a fear response, they can respond in a manner that manages fear or in a manner that allows fear to manage them.
  - ◊ Managing fear is productive and contributes to goal-setting, personal growth, and moving along a continuum toward self-actualization.
  - ◊ When a person is managed by fear, the results can be debilitating, paralyzing, counterproductive, and can adversely impact growth and development.
- Decisions to act are either causal or corollary:
  - ◊ *Causal* responses refer to cause-and-effect scenarios and short-term thinking most often linked to survival instincts. They are generally one action, one outcome.
  - ◊ *Corollary* responses are higher-level processes that require longer-term planning in an effort to increase one's probability of achieving a goal or intended outcome. There are usually multiple actions, over a period of time, that build toward a desired outcome
- Reliance on self (self-determination) and/or on others (dependence) also plays a key role in managing fear.

For more information about the METUS Principle, please refer to *The METUS Principle: Recognizing, Understanding, and Managing Fear*.

# Fear, Development, and Mathematics

## Fear

Can the METUS Principle be used to explain why so many students dislike math or have high levels of anxiety related to math? The answer is yes. The METUS Principle can explain fear caused by math and can also support those who experience math-related fear in the following ways:

1. If and when a person is able to recognize the role that fear plays in the decision-making process, he or she can begin a path toward empowerment. The person can learn to understand fear and how it impacts choices the person makes. When people learn to both recognize and understand fear, they can develop strategies for managing fear in a way that is consistent with behaviors and goals.

    As it relates to this book, these behaviors and goals are specific to math achievement. As it relates to education, and specifically math achievement, we will address the importance of understanding fear as it pertains to students.

2. If and when individuals are able to recognize the role fear plays in the decision-making process, they can develop strategies that motivate others to change their behavior(s).

As it relates to education, and specifically math achievement, we will address the importance of understanding fear as it pertains to teachers and teaching within our education system.

The key to achieving goals—and achievement in general—can be viewed as a three-part system:

1. ***Step One*** is learning to recognize the fear and self-limiting beliefs that create barriers.
2. Once you are able to recognize your fear, ***Step Two*** is to understand it. Understanding fear requires reflective learning. How has the fear of math contributed to your lack of success as a math student? What is it about math that you fear or dislike? As a teacher, what are you doing to make students less fearful of math and help reduce anxiety that is often felt by students in a math classroom?
3. ***Step Three*** is to manage or even embrace that fear and move forward rather than be paralyzed into non-action or give up on goals. It is incumbent upon us as thought leaders to create environments where students feel welcomed and supported—and in time, provide a place where students *want* to be, rather than a place they *have* to be.

We encourage you to take a moment to reflect on successes and failures in your own life. In almost all cases, we would be willing to bet your end results could be traced to whether or not you were able to manage your fear, or you allowed your fear to manage you.

The METUS Principle was developed to help you and others learn to recognize fear in its many disguises, to better understand the source of those fears, and then to manage those fears. As education professionals, it is essential we create environments for students in which they appreciate, understand, and develop mathematical aptitude—rather than fear and avoid it.

Unfortunately, for many of our students, math classes haven't provided them with a lot of positive experiences. As a result, over time, fear and anxiety are often associated with math.

## How Fear and Anxiety Impact Learning

Whether you are an education professional or a parent, it is important to recognize that school can be a major source of fear and/or anxiety in a child's life. A student's fear may be tied to academic performance—high achievement in addition to low achievement, as both can be connected to self-esteem.

Many high achievers place pressure on themselves, or perceive pressure from others, to maintain a high academic standard. Low achievers can face personal embarrassment and/or frustration and may be required to do additional work—all of which can be a source of anxiety or fear.

It is both counter-intuitive and counter-productive to provide students with more of what they are not good at and do not like without offering a drastically different approach. If, as a teacher, you are not using different strategies to support students who struggle, then it should be expected that more of the same actions will achieve more of the same poor results.

> **TEACHERS**
> Have you ever stopped to consider that when students act out in your class, their negative behaviors may be a call for help? What about a desire to be heard? While the attention they receive may be negative, it is nevertheless *attention*.

Both high and low achievers may also become targets of bullying, which impacts how safe they feel in their environments and whether or not they feel a sense of love and belonging—the lack of which can damage self-esteem.

Teachers and other education professionals can also experience anxiety and fear. Whether it be job security, job performance, or wondering whether or not they are impacting each and every one of their students, there are

> **ADMINISTRATORS**
> Thanks to increasing involvement by federal and state governments, one can reasonably argue that public education is driven by compliance more so than any time in history. With all the mandates forced on our teachers, they are constantly being asked to DO MORE WITH LESS.

# Fear, Development, and Mathematics

> **Parents**
> Have you ever thought about how your own personal views and the words you use influence your children? If your sons or daughters hear you talk about not liking math or hear you talk about how you don't use the kind of math being taught in school, how can you a) expect your children to appreciate math and give their best effort, and b) persevere at times when they may struggle?

pressures to perform. Often, a teacher's comfort level and his/her level of job security will have an impact on students and how comfortable they are made to feel in the classroom. Teachers who have high levels of anxiety are often less patient, less understanding, and spend less time getting to know their students because they are preoccupied with their own struggles.

The classic *nature* versus *nurture* debate is relevant and worthy of consideration as it relates to academic fear. School systems often face challenges specific to their geography and student demographics. This should not be used as an excuse but rather accepted as a reality. Kids in inner-city urban environments, versus rural environments, versus the suburbs, have unique environmental factors that can support or mitigate their fears in both the short and long term.

Once we recognize and accept that experiences are unique to each child (albeit similar as a result of their communities and home environments), we can help children progress toward their ideal, self-actualized selves.

We can help them recognize, understand, and manage personal fears that are holding them back and keeping them from being productive or successful math learners. If we allow ourselves to view student behaviors through this alternative lens, we won't necessarily see a child as a problem—but rather see a child who is afraid and/or unable to manage his or her fear(s) effectively.

If we can shift our way of thinking in order to help students become the best versions of what *they* want to be, rather than someone we *expect* them to be, and if we can help them understand that certain choices lead to more positive outcomes for them, we can change our communities in a positive way—one child at a time. This is student-centered learning, and as an education system we need to continue to promote classrooms that are student-centered as opposed to teacher-centered.

Similar to students, no two teachers are alike. Each teacher has his or her unique life experiences prior to becoming an educator. A teacher is a product of his or her unique personality, experiences, and education and relates to different students differently based on his/her own uniqueness. Teachers have their own

> Students have no control as to whether or not they were born into families of affluence or poverty. Students have no control as to whether or not their parents value math (or even value education, for that matter). Students have no control over the direction of the prevailing political winds that constantly impact the education they are receiving in some way, shape, or form.
>
> Students do have control over one thing, and this one thing is powerful! They have control over their effort. Helping students learn to recognize, understand, and manage their fears will help them become more engaged, more productive, and in time, more successful as learners.

fears associated with teaching, and each teacher manages those fears differently.

Whether you are an administrator, teacher, parent, or student, we believe it is important to understand fear and recognize how it impacts an academic environment. By recognizing, understanding, and managing fear, it is possible to create a positive and productive learning environment—an environment where fear is managed so that students and teachers alike can work on moving closer to their own versions of their ideal selves.

## How Fear Works

Understanding how fear affects the body, body chemistry, and the mind is profoundly important. Fear is the cognitive and physiological response to a perceived threat. It's a survival mechanism that signals our bodies for a fight, flight, or freeze reaction. True fear is essential in order to keep us safe. However, people who live in constant fear are at risk of becoming incapacitated.

Fear prepares us to react to danger, whether it's real or perceived. Once we sense potential danger, our bodies release hormones that slow or shut down functions not needed for survival (such as our digestive systems) and sharpen functions that might help us survive (such as eyesight and other senses). Our heart rates increase, and blood flows to muscles so we can run faster. Our bodies also increase the flow of hormones to an area of the brain known as the amygdala in order to help us focus on the presenting danger, which is then stored in our memories.

## Fear's Effect on Thinking

Once fear pathways are ramped up, the brain short-circuits more rational processing paths and reacts immediately to signals from the amygdala. In this overactive state, the brain perceives events as negative and remembers them that way.

The brain also stores all the details surrounding the danger—the sights, sounds, odors, time of day, weather, and so forth. These memories tend to be very lasting, although they may be fragmented.

Later, the sights, sounds, and other contextual details of the event can become stimuli themselves and trigger fear. They may bring back the memory of the fearful event, or they may cause us to feel afraid without consciously knowing why.

Because these cues are associated with previous danger, the brain may see them as a predictor of threat. This often happens with post-traumatic stress disorder (PTSD). For example, a soldier who experienced a bombing on a foggy day might find himself panicking when the weather turns foggy—without knowing why.

Think about students who have less-than-desirable home environments. Think about students who struggle socially, academically, or both.

Now, think about how traumatic environments in a student's life may contribute to, and even perpetuate, chronic failure.

### FEAR AND ANXIETY AND THEIR EFFECTS ON DEVELOPMENT

As cited in *The METUS Principle* and Paul Tough's book, *How Children Succeed: Grit, Curiosity, and the Hidden Power of Character*, there is significant medical research and a number of examples that highlight the impacts fear and stress have on people—including their learning.

Exposure to chronic stress, especially early in life during a child's physical and cognitive development stages, can produce serious and potentially damaging negative effects. Because of the chemical reactions in the brain, overactive and sustained activity by the hypothalamic-pituitary-adrenal (HPA) axis can adversely affect physical,

> **METUS:** We see the effects of fear in student outcomes every day. Students are unable to focus on completing higher level tasks because their minds are preoccupied. Students living with chronic fear are focused in survival mode, and when you are trying to survive, it is very difficult to thrive.
>
> *Causal (short-term) vs. Corollary (long-term) thought patterns*

psychological, and neurological development—particularly in children.

While our school systems cannot control what environment a child experiences off campus, they can play a significant role while a child is in school. Our schools and our classrooms need to be safe havens for children and provide the environment and relationships that help reduce and limit fear.

In his article, "Brain on Stress: How the Social Environment Gets under the Skin," (*Proceedings of the National Academy of Sciences*, Oct. 2012) Bruce McEwen, a neuroendocrinologist at Rockefeller University, has proposed a theory of managing stress he calls "allostasis." According to McEwen, allostasis is a process that causes wear and tear on the body. When the body remains in a constant state of stress, or the body's stress management system is overworked, eventually the body encounters problems as a result of the strain. He calls these problems "allostatic load," whereby the effects of constant stress (fear) can be seen physically throughout a person's body in the breakdown of cells and the inability of the body to protect, heal, or recover from damage.

In addition to McEwen's work, stress physiologists have found biological evidence that stress likely alters parts of the brain. This, in turn, results in changes with regard to how a person processes information. The prefrontal cortex is the part of the brain most affected by early stress, which is critical in self-regulatory activities of all kinds—both emotional and cognitive. Studies have shown that children who grow up in stressful environments are more likely to have problems concentrating, sitting still, and recovering from disappointing experiences (*Stress- and Allostatsis-Induced Brain Plasticity*, Researchgate.net).

> We all know children in difficult environments, but do we truly and earnestly appreciate their struggles and how they impact their ability to learn?
>
> As an administrator or teacher, have you ever considered the effects that chronic fear and anxiety have on learning? What are some ways that you have personally attempted to help a student address his or her fear?

> **TEACHERS & ADMINISTRATORS:**
> Do you find that students who come to school from stressful environments are more likely to act impulsively and exhibit more maladaptive behaviors than students who do not? What are some strategies that your district, school, or classrooms can do to try and reduce anxiety and fear for students who need the most support?

At the base of Abraham Maslow's needs pyramid are physiological needs (food, water, shelter, etc.), then safety needs (safety and security), and then needs for love and belonging. It is very unlikely that a child in a stressful environment is able to have all three of these important levels of need satisfied.

Whether it is a lack of food or clean water, living in an unsafe environment, or growing up in a situation devoid of love and support, young children do not have the mental maturity to recognize how fearful they are. They are unable to make sense of their environments or understand their fears. As a result, they are unable to manage their fears.

**MASLOW'S HIERARCHY OF NEEDS**

Children in fear live moment to moment and develop behaviors and mental thought processes that heavily emphasize causal thinking tendencies—in other words, "What do I actually need to do *right now?*" Because they are overwhelmed, they are not able to spend much effort developing corollary thinking skills or be able to think ahead to alternative futures based on a series of choices. In other words, students in a state of chronic stress operate in, and are stuck in, survival mode.

When children fail to develop corollary thinking skills, the brain wires itself along accustomed pathways. As a result, over time, such children's brains become wired in a way that they struggle to control impulses, are distracted by negative feelings, and retain fear. These are all factors that negatively influence learning.

Once more basic needs are met, a person will begin to think beyond the "here and now" and develop brain pathways that support corollary or long-term thought processes, which in turn positively impact learning.

As far as brain development is concerned, some effects of stress in the prefrontal cortex can be described as emotional or psychological, which can cause anxiety and depression.

Lawrence Steinburg, a psychologist at Temple University, has recognized two separate but related neurological systems that develop in childhood and adolescence and together have a profound effect on the decision-making patterns and habits of adolescents.

It is in early childhood that our brains and bodies are most sensitive to the effects of stress and trauma, but it is in adolescence that the damage stress inflicts can lead to the most serious and long-lasting problems.

According to Steinburg, the first neurological system is known as the "incentive processing system." This system makes a person more sensation-seeking, emotionally reactive, and more attentive to social information.

The second is called the "cognitive control system." This second system allows a person to regulate the urges of the incentive system.

From the perspective of the METUS Principle, the incentive processing system is the system that supports a causal thought process: "If I do [ACTION], the outcome will be [RESULT]." This is simple cause-and-effect thinking and does not require much higher-level cognition. Students operating in this mode are not thinking about the future; they are simply trying to survive day to day.

The cognitive control system, on the other hand, is more refined. It takes into account a person's immediate impulse to do something, and it regulates behaviors by assessing a number of variables, goals, and a range of possible outcomes. The cognitive control system allows a person to think in a corollary fashion: "If I do this or that, these are the possible outcomes associated with possible choice(s) I could make." This is the type of thinking required to establish effective career- and college-ready environments.

Steinburg suggests that the incentive processing system peaks in early adolescence, and the cognitive control system doesn't finish developing until a person reaches early adulthood. However, just because the two systems don't develop until certain phases in life, this does not mean they entirely lack any ability to process information earlier in life. A young child may have some capabilities of each, and the extent to which a child will process information will vary.

These are but a few of the many studies that demonstrate the negative effects that stress plays in our lives. The more stress we have, especially early in life, the more primitively we think and act. The less fear we have, the better we develop, and are able to develop, our higher-level thought processes.

Experiencing self-actualization, and becoming self-actualized, is a continuum or progression. It takes time and requires us to progress from the person we are to the person we want to become—our own personal versions of our ideal selves.

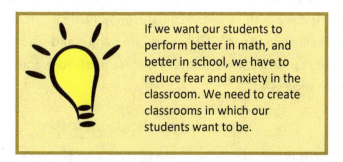

If we want our students to perform better in math, and better in school, we have to reduce fear and anxiety in the classroom. We need to create classrooms in which our students want to be.

If individuals only develop causal thinking skills, they will be unable to appreciate the future and envision their ideal selves. If they cannot envision their ideal selves, they will not be able to develop the corollary thinking skills needed to help them progress from the person they are to the person they want to become. As a result, a child will not value learning because it has little value in his or her day-to-day life.

# DEVELOPMENT

### THE OTHER END OF THE SPECTRUM

For those of you who find it difficult to understand the METUS Principle, we ask that you consider research done by Shawn Achor in *The Happiness Advantage*. Achor's approach to success comes from the other side of the same coin, but the end result is the same.

In the METUS Principle, the less fear and anxiety we have, the more likely it is that we will achieve our goals.

Achor asserts that when we develop a positive attitude and find joy in the work we do, the more likely it is that we will achieve our goals. "When we are happy, and when our mood is positive, we are smarter, more motivated, and thus more successful."

Both approaches mean that students who become *happy* learners can become productive and *successful* learners.

But what exactly is happiness? According to Achor, happiness is defined as an experience of positive emotions combined with deeper feelings of meaning and purpose. Happiness implies a positive mood in the present and a positive outlook for the future. When assessing happiness, we can look at three measurable components: pleasure, engagement, and meaning.

Aristotle referred to happiness as *eudaimonia*. This roughly translates to "human flourishing." In other words, happiness is the joy that we feel striving after our potential.

The key to helping students perform better in math, and helping teachers teach math better, is to help all involved find joy and happiness in their daily work. However, from a practical point of view, who can possibly find happiness in skill-and-drill activities? There is only so much (or so little) joy to be found in memorizing formulas or solving step-by-step equations. Instead, the goal is for happiness to be found in the relationships, discussions, rich activities, and in those "aha!" moments experienced in the classroom.

Happiness in the classroom also translates into good student attendance. Students who find joy in school, their classes, and their work, are more likely to maintain good attendance. As teachers, this means that we should be

> Research shows that unhappy employees take more sick days, staying at home an average of 1.25 more days per month, or 15 extra sick days per year.

striving to create environments that promote happiness and are an inviting place for students to come every day. Needless to say, it's hard to effectively engage students if they are frequently absent.

> **TEACHERS:**
> Keep these three components in mind when considering a productive math classroom:
> - Pleasure
> - Engagement
> - Meaning

> **TEACHERS:**
> Do students in your classroom feel like they are flourishing—or merely surviving?

## DEVELOPING A LEARNER'S MINDSET

Results are important. However, we have to remember that results are cumulative and a product of our daily work.

Before results can be seen, we have to be willing to embrace small wins and help students develop a *learning mindset*. A learning mindset is an aptitude for learning that requires zero talent, but when focused and applied helps students to develop good habits that foster learning. Traits within the learning mindset include:

- Being on time
- Having a good work ethic (giving your best effort)
- Using positive body language (focused attention)
- Maintaining good energy (not quitting)
- Having a good attitude
- Having passion
- Being coachable
- Doing extra (going the extra mile)
- Being prepared

As learners, when we actively engage, we will learn. In some cases, learning may appear to be easier for some than others, but we should do our best to prevent students from falling into the trap of comparing themselves to others. When we can help students better understand the importance of focusing on the learning process, they develop an appreciation for learning as a means to achieving their goals. Over time, when they learn to develop the personal traits that contribute to a learning mindset, they are well on their way to developing their talents.

We as teachers and professionals need to do a much better job at making math "real." Math is more than numbers and equations, and it needs to be presented as such. Math tells stories. We need to help students learn to approach math as detectives in order to answer questions about what has happened or what has been. Students can

also use math to be creative and to create. Bringing math to life allows students to enjoy their work and experience meaningful and fun learning.

## TEACHING VS. LEARNING

Far too often, we find that teachers believe they need to *control* instruction. They don't. Rather, teachers need to *inspire* students to build upon their intuition, establish meaning, and develop understanding. In doing so, learning happens naturally.

The reason we wrote this book is because we believe that all students are not only capable of learning math, but

We need to create environments that attract students and are engaging. We need to create classrooms in which students *want* to be, as opposed to those they *have* to be. A positive, engaging, collaborative environment that is student-centered, reduces fear and, in many cases, produces happier kids. In today's innovation-driven knowledge economy, business success in practically every job hinges on being able to find creative and novel solutions to problems. We should be ncouraging and fostering this type of learning in our math classrooms and not falling into the process, procedure, and memorization rut.

> Your classroom environment matters, expectations matter, rapport with students matters, and by focusing on creating the right environment, you establish a culture that produces the desired outcomes and results.
>
> To get quality results, we need to focus on helping students develop a learner's mindset so that they develop conceptual understanding through positive engagement.

also enjoying math under the proper conditions. We also believe that all teachers, from elementary through graduate school, are capable of establishing the right conditions for students to appreciate math—as well as the right conditions for teachers to thrive and enjoy teaching.

*Teaching* is a one-way street that only requires a single individual—the presenter. *Learning*, however, is a two-way street. Learning requires both a presenter of information as well as a receiver and processor of information. The most effective way for a presenter to engage a receiver is by helping him or her establish a connection between the heart and the mind.

> Learning is a two-way street. Do you prefer to have people talk AT you, or do you prefer to have people interact WITH you?

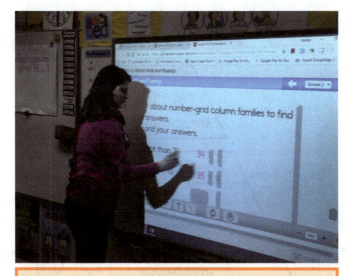

**Teachers**
Do you spend the majority of your time in class presenting information? If you are spending much of your time talking, with your back to your students, how engaged do you think they really are?

When a connection is made between the heart and mind, magic happens! What magic, you might ask? The answer is a reduction in fear and anxiety, and the capacity to embrace and apply information and ideas. The following pyramid, modified from Abraham Maslow's needs pyramid and taken from the framework of the METUS Principle, provides a visual representation of the type of learning environment that is necessary for learning.

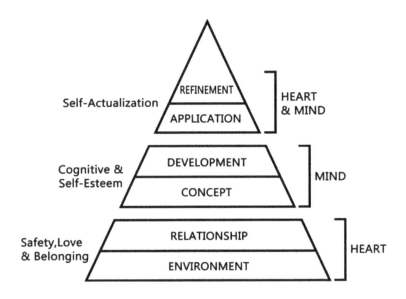

## LEVEL I: HEART

At the foundation, learning requires a positive environment and positive relationships. Students need to feel safe, loved, that they belong, and that they matter. When students truly believe their teachers care about them, have their best interests in mind, and genuinely want to see them succeed—not only in the classroom, but in life—students are able to allow themselves to be vulnerable and open their hearts to learning.

## LEVEL 2: MIND

Once teachers help to draw a connection to the heart, they can begin to challenge the mind. The more we learn, the more we grow as individuals.

When we help students learn new information, create concepts, and develop conceptual understanding, we are also helping them with essential cognitive and self-esteem needs. The feeling of accomplishment, self-worth, and potential are powerful emotions that stimulate the mind.

## LEVEL 3: HEART AND MIND

The pinnacle of the learning process is created through a synergy of the heart and the mind. It occurs when students are able to appreciate how their efforts will help them achieve personal and professional goals. It is a point at which they seek to apply and refine knowledge in a very personal way and no longer require external motivation for learning to take place.

Self-actualized learners desire to apply and refine information; they are self-driven and self-motivated because they understand, appreciate, and value knowledge as a means of sustaining that synergy of heart and mind.

## Motivation, Goal Setting, and Growth

According to Maslow, and as demonstrated by the METUS Principle, once basic needs such as food, water, and safety are satisfied, humans are naturally motivated to continue to embark on an ongoing quest in order to reach their full potential.

In education, we often hear the topic of motivation tossed around. "Oh, Bobby is just not motivated," or "Kendra doesn't see the value in [fill in the blank] so she doesn't pay attention in class." Is it truly a lack of motivation or disinterest, or is there more to it than that?

As leaders, and when working with students, we need to appreciate the uniqueness of our students. Every student, at any given time, is experiencing a personal struggle; though of course, degrees of struggle will vary from child to child. Furthermore, perception is reality. Whether or not we believe a struggle to be "important" doesn't mean that it isn't important to the child who is experiencing it. We must appreciate the struggles of others and be supportive to establish rapport and gain trust.

## Learner Profiles

Differentiation in education requires teachers to tailor instruction to meet individual student needs. In some cases, differentiation requires a teacher to differentiate content. However, more often differentiation means presenting

> If you want to help students reach their potential and achieve success, merely presenting information to them isn't enough. You need to engage their hearts and their minds and help them take ownership of the learning process. This requires moving from the presentation of information to the facilitation of learning.

information, or even a line of questioning, to students in a way in which they can relate or connect. Making personal connections for and with students isn't easy, but it is important.

The following describe three learner profiles of struggling students to help you better establish relationships and improve student productivity in your classroom.

- **Profile #1**—Motivated, but with limited resources
- **Profile #2**—Motivated, but inconsistently
- **Profile #3**—Motivated by external rewards and/or fear, but only temporarily

### #1—Motivated, but with limited resources

If you have successfully connected with your struggling students, and they are telling you that they are willing to try making changes, help them build upon that motivation. When it comes to changes in behaviors, the first hurdle is motivation, but after that, it comes down to "ability," according to social scientist, B.J. Fogg.

Fogg states that we all experience, to varying degrees, scarcity in resources such as time, money, and skills—and that scarcity can interfere with our ability to accomplish even what we are sufficiently motivated to achieve.

> According to Fogg's behavior model, a person has two options:
>
> 1. The person can try to get more of the resource that he/she is lacking (easier said than done) or
> 2. The person can scale down the behavior to match the resource that is available (more practical).
>
>    Want to start studying but find it difficult to sit down and focus for 30 minutes at a time? Start with 5 minutes.
>
>    Want to get better grades but have no idea where to start? Find a friend who is successful, and start asking questions.
>
> Source: *A Behavior Model for Persuasive Design*, www.mebook.se

### #2—Motivatied, but inconsistently

Motivation waxes and wanes. Be ready with options.

In the throes of inspiration, students often set ambitious goals that seem entirely doable to their highly motivated selves. "Starting today, I am going to get straight A's in school!" Although this might be a wonderful goal, the minute

> *Take a more challenging path when you are feeling inspired and an easier route when motivation is waning.*

they fail to meet these high expectations, some students throw in the towel. They quickly go from riding a high of ambition to a low of, "I guess I will never be a straight-A student."

We, as teachers, often forget that motivation isn't constant. Sometimes students are just not "feeling it," and other times, they simply fall short despite a strong effort, so it's important to build in daily options to harness their "motivation wave." In many cases, daily or even hourly fluctuations in motivation can occur depending on the class, the teacher, the assignment, and so on.

---

**FOR THE STUDENT:**
**TIP: DO YOUR HOMEWORK BEFORE SOCIAL MEDIA OR OTHER PERSONAL ACTIVITIES**

Let's say your goal is to complete one page of math homework every night before bed. You get home late from an event one night and just want to roll into bed. Your motivation wave is hitting bottom. Instead of blowing off your new habit completely, make it easier for yourself and start with a few problems. You may still cut your homework session short, but the fact that you adhered to making an attempt will reduce the chances of falling off the wagon entirely.

### #3—Motivated by external rewards and/or fear, but only temporarily

Have you ever had a flash of motivation upon learning a sobering, new fact? Maybe you read that excessive sitting can lead to diabetes, so you suddenly bolt out of your chair every hour. Or your school launches a "biggest loser" competition for teachers that includes cool prizes—so you immediately start skipping meals in an attempt to drop 10 pounds fast.

Your mission succeeds—for a few days.

The fact is, change inspired by fear or external rewards rarely lasts. This is as true for adults as it is for students. Simply rewarding them or punishing them will only take them so far before they reach the point of diminishing returns—or even worse—quitting.

It can be inspiring to be tempted with a prize, but rather than create an environment in which students focus on a short-term goal, help them relish the positive experience of learning something new. If your students do not develop intrinsically motivating reasons for taking on a new habit, and striving toward goals that are personally relevant or meaningful, then they won't retain this new behavior as a part of their daily routines for long.

### *Identifying the "WHY" is motivating*

When the going gets rough—when good intentions go up against ingrained behavioral patterns—knowing and remembering what's really driving us (the WHY) may be all that keeps us on course.

Knowing the WHY is immensely important for teachers, but it's equally important for students. Simply being in class or in school because it is a requirement or prerequisite for kids ages 5 to 18 is not a WHY.

The technique therapists and coaches use to uncover the WHY is called "motivational interviewing," and it's something everyone can borrow and use to help them focus on a particular task. As a teacher, if you notice students disengage or begin to check out, help them connect with their WHY until the answer gets real and they see the value in what it is that you are covering in class.

Here's a conversation you might have with your student:

> **Teacher:** "Why are you in school?"
>
> **Student:** "Because I have to be. If I didn't, I wouldn't be here!"
>
> **Teacher:** "OK, fair enough. Why do you think that there are laws that require you to be in school?"
>
> **Student:** "Because! I dunno, man. Who cares?"
>
> **Teacher:** "I care, and the system cares. Someday, maybe even today, you will have to look out for

yourself. You will have to find money for rent, pay the bills, eat ... which is a lot harder if you don't have an education. I don't want you to struggle; too many people out there have to struggle. I want better for you. Do you understand what I'm saying?"

**Student:** "Yeah, I guess so."

**Teacher:** "So tell me, what types of things are you interested in?

**Student:** "Sports, movies, music, you know..."

**Teacher:** "Tell you what. I will try to make some of this information relatable to you and use sports, movies, or music to make it interesting for you."

*Is this exchange an oversimplification? Perhaps. But conceptually, the argument is valid. As educators, we need to take a greater interest in our students to engage them and help them develop as learners.*

### Taking "WHY" one step further

As human beings, we struggle with concepts, ideas, and beliefs that aren't relatable or personally relevant. If something doesn't interest us, at the very best, we tune it out. There are some people who resent, openly dislike, or even express hate for things they do not understand—and the same is true for math.

When it comes to math, so much of what is taught focuses on processes, procedures, formulas, and equations. This can sound like mere "blah, blah, blah" to the students.

# Fear, Development, and Mathematics

If they cannot relate to the math, all the talk surrounding math becomes noise.

The funny thing is, MATH IS ALL AROUND US. Quite honestly, and literally, you can take just about any interest students have and *show them the math*.

Showing students the math has value; helping them relate to the math has value; getting students to open their minds to WHY math is useful and helpful to them has value; and you know how we do this? We put math in their hands!

Once math is felt, it can then be understood. It is much harder to wrap our minds around the math if we are not putting our hands on or in it. We often hear, "We need to move from the concrete to the abstract." How many teachers truly understand what that statement means?

In our world, concrete can be felt, lifted, thrown around, and broken apart. We need to have math in our hands so that in time, we will feel it in our hearts and in our minds.

If students are just memorizing formulas and algorithms, they are simply becoming math robots. For some students that works fine; however, we believe we can and should be doing more than just creating receptacles of information.

There are many teachers who can tell us what and how—but far fewer who can break down a subject and show someone WHY. There are even fewer who bring math to life. We challenge you—be one of *those* teachers who brings math to life for your students!

# I'm Just Not a Math Person!

> The METUS Principle is about breaking down and managing fear in manageable parts. Through managing fear at each level of the hierarchy, an individual is able to move toward his or her "ideal self." As we help students address their fears and attend to their needs, they become more productive learners.

## Questions to Consider:

- How did *you* learn math?
- Was your experience student-centered?
- If so, what did that look like, sound like, and feel like?
- If not, how would you describe what we should be providing for students now?
- Teachers, what can you do to help students feel comfortable and supported?
- Can you think of an example when one of your students performed better because you went the extra mile to help support him or her? How did the student's success make you feel?

As teachers, we all encounter different challenges with different students on different days for different reasons. It is important that we maintain a caring, student-centered environment in which students feel comfortable and supported and continue to help all students progress in a productive manner.

# MATHEMATICS

## The Practitioner's Lens

> *"You can lead a horse to water, but you can't make it drink."*

This was a statement once made by a teacher when expressing his frustration teaching kids who don't want to be in the classroom (and kids who seemingly aren't interested in learning).

This statement isn't new, and it is probably something most of us have said at one time or another, for some reason or another. On its surface, it makes perfect sense. We can put water right in front of a horse's mouth, but it is up to the horse whether or not it drinks. The premise holds true in the world of education as well. We can present information to our students, but we cannot make them learn.

So what? You can't make a horse drink and you can't make a child learn. No profound epiphany here.

Let's consider our horse once again. True, you cannot make it drink. However, you can make it thirsty. Put a block of salt in front of it, and the horse will lick it. Then it will get thirsty, and then it will drink!

Now think about children. You cannot make students learn simply by putting information in front of them. What can you do as a teacher to make your students thirsty? What can you do to engage them and provide the right environment so they *want* to learn?

## Q: WHY DO PEOPLE HATE MATH?
## A: FEAR

### RECOGNIZE

According to an AP News Poll, 4 out of 10 Americans hate math. ("The Most Unpopular School Subject: Poll Shows Love-hate Relationship with Math," Aug. 17, 2005). We conducted our own study, whereby we asked 1,000 individuals a single question: *"Do you see yourself as a math person?"* Individuals from children to adults, across five Midwestern states, were presented with three options: YES, NO, or NOT REALLY.

Of those asked, 817 responded with either NO or NOT REALLY, showing us that 4 out of 5 individuals feel a disconnect with math. Only 183 individuals responded by affirming that yes, they consider themselves to be math people.

The results from both AP and our own survey confirm that for many students, teachers, and parents, math can be a source of anxiety and a subject many Americans do not readily embrace.

# Fear, Development, and Mathematics

> Children pick up on the fears or dislikes of their parents. Parents who openly speak negatively about a particular subject, or learning in general, do their children a disservice. Do not project, as parents, your own struggles or dislikes onto your children. Give them a chance.

Fear causes people to respond in a number of ways, but common responses include fight, flight, or freeze. This is also true when considering how people view or relate to math. Some people will sit silently and not tell anyone they are fearful. Others will loudly voice their disconnect with math through comments like, "I just wasn't born with the math gene!" Some people will get defensive or mad when they feel their intelligence is being challenged. It is important to recognize who is fearful and understand their responses.

## Understand

To improve our education system, it helps to understand why, how, and where a system is deficient or needs improvement. We spend a lot of time supporting administrators, teachers, parents, and students in the area of mathematics achievement. Significant attention is paid to the culture in a community, district, school, and even narrowed down to a single classroom.

One of the questions we are often asked is, "Why do people hate math?" Our notes in regard to that answer have less to do with math itself and more to do with how math is taught—and furthermore, how math is taught relative to other subjects. Math often creates anxiety and is feared because:

- It requires precision, generally perceived as having "only one right answer," and
- It requires attention to detail, generally perceived as having to follow defined steps (process and procedure) in order to solve an equation.

With math, it is harder to mask shortcomings, whereas subjects such as reading, writing, and social sciences:

- Favor creativity and expression over precision (big ideas vs. details), and
- Are more subjective and open to interpretation, and are more accepting of errors.

Reading, for example, offers contextual clues. If a word is unfamiliar or challenging, students can often skip words and still understand a story or concept about which they are reading. To write a book report, students can buy *Cliffs Notes*, read online Google summaries, watch the movie, or get a summary from a friend. It is more difficult to cover up or hide a lack of effort or understanding when it comes to math, because there are no easy shortcuts.

Lastly, consider the draft and review process. In writing, we encourage trial and error. How many authors sit down, crank out a book, and consider it done? The answer is—ZERO. When writing, we draft and review, draft and review, draft and review. We solicit advice from others, review recommendations, and revisit our original work. We not only accept a critique of our work, we expect and encourage it.

Why is it that we allow creativity, exploration, and multiple attempts for other subjects, but we expect precision and perfection when it comes to math? In other words, why isn't math allowed to be a "messy" process, too?

We contend that if society demanded precision from all writers and adhered firmly to spelling, grammar, and punctuation rules in the same way that we require precision and exactness in math, we would see students fear literacy as much as they do math.

In other words, the notion of being or not being a "math person" is socially constructed, which presents both advantages and challenges to learning. One important advantage to recognizing that the fear of math is socially constructed is that it disqualifies a person from attributing success or failure in math to the almighty "math gene"—something we are born with or without.

ALL students are capable of learning mathematics, but it requires hard work from all parties involved—administrators, teachers, parents, and students. In her book,

*Mindset, the New Psychology of Success*, Carol Dweck has written about a student's mindset and how that can affect learning. "Students' theories of what it means to be intelligent can affect their performance. Research shows that students who think that intelligence is a fixed entity are more likely to be performance oriented than learning oriented—they want to look good rather than risk making mistakes while learning. These students are especially likely to bail out when tasks become difficult. In contrast, students who think that intelligence is malleable are more willing to struggle with challenging tasks; they are more comfortable taking risks." As a community, we need to support a growth mindset for learning mathematics. As we help students persevere and work hard, they will learn.

# FEAR, DEVELOPMENT, AND MATHEMATICS

## MANAGE

### The Value of Precision

When teachers, students, and parents hear the word "precision," they often associate precision with perfection—or similarly, exactness. However, precision is *not* perfection. We are not, nor should we be, demanding perfection from students when it comes to math.

Teachers typically know one method for solving a problem. Therefore, they limit the students' creativity in finding their own methods or pathways to a solution. But what if precision didn't focus on a single pathway? What if teaching math emphasized the importance of learning processes in a way that was meaningful for students with attention to precision? This process at first may be challenging as students attempt to make sense of problems and develop an understanding of their problem-solving process. As students practice processing information, however, with teacher support they will develop greater precision.

### KRIS'S CLASSROOM

My son was working on his online math homework, but he wasn't sure how to solve the task. I told him to click on the worked example so that we could read it together and try to figure it out. I was shocked to see that the software told him to "plug in" the number to solve for x.

I plug in my phone, curling iron, or hair dryer—we do not "plug in" in mathematics. *Substitution* is the mathematically precise word for "plug in." I did not expect this slang from a mathematics publisher.

As my son continued working, the next task asked for the x-intercepts. He typed in 1, 3 and it was marked correct. This was another example of the software not teaching students to attend to precision. The x-intercepts are really "(1, 0) and (3, 0)"—using parentheses.

When writing coordinates in math, we use the form (x, y). Therefore, (1, 3) means that x equals one and y equals 3. If students are allowed to write 1, 3, they are not connecting the numbers to the true meaning of the intercepts—which is where the graph crosses the x-axis.

When we *consistently* ask students to write the coordinates (1,0) and (3,0), we are reinforcing that the graph will cross the x-axis and the y values are zero. When we wonder why our children are not making connections, consider how we (and/or the software they're using) are attending to precision.

> **Standard for Math Practice 6:** Attend to precision
> *(see more in Addendum)*

## Vocabulary

**Intercepts:** Place where the graph crosses the x or y axis

**Axis:** The vertical and horizontal lines that make up the quadrants of a coordinate plane. The vertical axis is usually referred to as the y axis and the horizontal axis is usually referred to as the x axis. Examples: Use the coordinate grid to plot the numbers on the x and y axis.

**Coordinates:** A two-dimensional surface by two intersecting and perpendicular number lines on which points are plotted and located by their x and y coordinates.

**Product:** The answer of an equation in which two or more variables are multiplied. Example: The answer to any multiplication problem.

**Quotient:** The number which is the result of dividing two numbers.

**Whole Numbers:** Also known as natural numbers. They are used to count the number of physical objects.

**Improper Fractions:** An improper fraction is a fraction where the numerator (the top number) is greater than or equal to the denominator (the bottom number). Example: 5/3 (five thirds) and 9/8 (nine eighths) are improper fractions.

Attending to precision in mathematics is a daily process and needs to be practiced and reinforced often. Teachers need to lead by example and model precision. In the classroom, a teacher will need to decide which moments require that attending to precision is non-negotiable.

When teachers are setting up their collaborative classrooms and beginning a new school year, they need to make sure that precision is discussed as a norm. I will describe this more later when I describe the first three days in my classroom.

Teachers must model mathematical precision, especially in language. I recently facilitated a professional development seminar for middle-school and high-school teachers. At the end of the five-day workshop, the teachers were asked to write a reflection paper answering the question, "How will I apply what I have learned in this workshop into my mathematics classroom?" One teacher reflected about precision in her classroom with this:

> "One of the most important lessons I took away from the academy was how crucial it is to use precise language and proper mathematical terminology in my lessons. For many years, I have used what I would now describe as "mathematical slang" in my classroom. While the students knew what I was talking about, when they took standardized tests, they sometimes did not recognize what they were being asked to do in a problem because I did not use the same terminology when I taught the lessons. This is something I have been working on and intend to continue to address in all future lessons."

# Fear, Development, and Mathematics

As teachers, we sometimes forget the impact our examples have on children every day. Now don't get me wrong—I want math to be FUN. In addition, I want students to remember and build upon what they know each day, week, and year.

As a teacher, I often put a "what" after students answer with a number, such as, "24 what?" As students start to put units and labels on numbers, they make more sense of the scenario. For example, my students tried to find out how much a tax consultant would make per hour if she worked "n" number of hours.

I overheard one student say to his group that the answer was $2,000. Another student spoke up and said, "You can hire *ME* for that job at $2,000 an hour. I wouldn't have to work much!" As the group discussed the context of the $2,000, it became apparent that something must be wrong. They revised their work to find the correct answer.

As we develop consistency, we help establish meaning—both applied and personal. As students practice math language, they will become less fearful of looking or sounding stupid.

I hear a lot of people say, "The children don't remember the math from one year to the next." I realized in my second year of teaching mathematics to 11th graders that they forgot a lot of information I knew I had taught them in Algebra I the year before. Even some of the students who received high grades in their previous math classes struggled to remember basic equation solving.

Why is this? We must practice math to maintain it. Exploring mathematical concepts and practicing the skills contribute to the ability of students to remember. I also believe that precision helps students' retention and connections from year to year as well.

When we ask students to be precise, we are also asking them to think about their work—to engage as reflective learners. Once I had students explore numbers in an elementary class, and I stuck register tape in six locations around the room. I then gave each group an envelope containing a set of fractions from 0 to 2½. The fractions were represented as whole numbers, improper fractions, proper fractions, and mixed numbers. I made sure that the length of the register tape was different for each group, and I told the groups they could not move the register tape.

Then I asked each group to create its own number line by placing fraction cards on the register tape. The students were not allowed to take the register tape off the wall or change it in any way. They were also told to be as precise as they could be.

I watched as the groups tackled this task in different ways. One group divided all the fractions up amongst the team, and everyone just started randomly paper-clipping their fractions to the register tape. After a minute, they stopped and decided to figure out who had the largest number and then taped that number onto the end of the register tape. They discussed how they would figure out where 1 and 2 were on their number line.

One student used her arm to measure. I then asked if there was a more precise way without using a ruler. Another student grabbed his notebook and started counting how many "notebooks" it would take to get to the end. As a teacher, this activity is about magnitude of fractions using benchmark fractions. I also needed to have the students attend to precision because fractions require equal parts.

Encourage students to be precise in mathematics, and you will see them begin to develop a clearer understanding of the subject. As students communicate precisely to others and express their answers with a degree of precision, specificity in their explanations will develop. They will also become more confident, which will lessen the fear of mathematics.

In elementary grades, students should practice formulating explanations regularly. By high school, this expression of explicit reasoning will be a natural part of their mathematics discussions.

Remember that we want students to become problem solvers. As students talk about mathematics using precise definitions, units of measurement, and labeling quantities, they will reflect on the numbers with which they are working. Some students and teachers see mathematics as getting the correct and/or numerical answer. As we shift our teaching focus away from the answer and more on exploring the concepts and procedures, students will find attending to precision natural.

Recently, I was observing in an Algebra II classroom and the students asked what "quotient" means. The teacher replied, "This is a fancy math term for dividing." Although partially true, this was a missed opportunity to clarify the definition of quotient and connect their understanding of sum, difference, and product—AND attend to precision with mathematical terms.

I would have asked the class, "What do we call the answer when we add? *Sum.* What do we call the answer when we subtract? *Difference.* What do we call the answer when we multiply? *Product.* In this quantity, the original logarithm is represented by division, therefore we call it *quotient* because the result produced by division is defined as a quotient."

Mathematical terms aren't just "fancy terms." They help students make connections with a basic understanding and number sense that they learn in elementary, middle-school, high-school, and post-high-school education.

Our hope is that you take time to reflect on your teaching practices and attend to precision. Take time to model precision in the classroom. Remember to support the students as they practice attending to precision.

A simple question like, "12 what?" will allow students to reflect on their answers and add units to their final answers. As students explain their reasoning, support their precision in language by asking clarifying questions. "What do you mean when you say...? What mathematical terms could we use that mean the same thing?"

Have fun with mathematics, and laugh. I loved it when the class used to mimic me by saying, "We plug in our phones; we substitute in math!" I smiled and said, "Great job attending to precision with your language!"

It is okay for the process of developing conceptual understanding and meaning to be a little "messy," as long as our efforts to teach concepts and terminology are precise. For example, entrepreneurial start-ups look very much like a messy, fail forward, collaborative, student-led classroom.

### SOME REFLECTION QUESTIONS

1. Why is math scary?
2. What does precision mean?
3. How are you supporting precision with mathematics at home?
4. Educators, how are you modeling precision in your classrooms?

# The Foundation for Meaningful Change in the Learning Environment

### Recognize

There has been, and remains, a belief within our society that some people are born with the "math gene," while others are not. For those who are not proficient in math, or have struggled with developing their math skills, participating in math classes or being required to demonstrate an understanding of math concepts can create anxiety and become a source of fear.

*Learning math should start at a young age through games and other engaging activities.*

## UNDERSTAND

Are some people born "math people" and others not? In 1869, Francis Galton officially began the age-old debate of "nature versus nurture." Since then, everyone from scientists to your average Joe and Jane have offered their perspectives. Do DNA and genotype predetermine who we are and what we are capable of, or are we born as blank slates? Does our environment dictate who we are and what we will make of ourselves by virtue of our experiences and interactions with the world around us?

Is there a math gene? The short answer is that to date, there remains no scientific evidence to support genetic advantages when it comes to intellectual capacity, aptitude, and/or advantages in particular subjects or content areas. In other words, there is no known math gene! However, there remains voluminous work that empirically and quantifiably demonstrates the power of learning environments and their ability to impact learners, both positively and negatively.

## MANAGE
### *A Positive, Student-Centered Approach*
### *Facilitation vs. Presentation*

It should be our goal as administrators, teachers, parents, and students of learning to make sense of mathematics. This means creating an environment that fosters learning rather than one that relies on the conformity of processes and procedures. To improve, we have to be willing to accept that the way we teach, the way we have been teaching for decades, and the traditional ways of learning need to change.

While we cannot control the environments in which our students are raised, we can certainly improve the conditions in which they learn. As education professionals in the

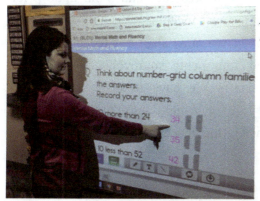

*A teacher can talk at students...*

*OR*

*A teacher can interact with students*

area of mathematics—teachers, instructional coaches, and administrators—we need to move students from their intuitive understanding of the real world into that of abstract mathematical representations and encourage them to develop their own understanding of mathematical concepts.

Students need to question why and how mathematics works in order to fully "own" their learning. Learners should notice patterns or structure and build upon prior knowledge in learning new concepts.

For this to happen, we must align our teaching to *how* students learn and stop being so myopic about telling students *what* to learn. We have to allow students to explore, discuss, critique, and make sense of the mathematics. The art of facilitating learning is the craft we call teaching.

To bring out the best in our students, we need to encourage the best from our teachers. If teachers are unwilling (the small few) or incapable of embracing such challenges, then it is incumbent upon leadership to seek those who are willing and able.

## The Foundation for Meaningful Change

### What are the First Steps in Creating a Student-Centered Classroom?

The first step in creating a student-centered classroom is establishing a safe environment in which all students feel they can talk, make mistakes, and learn without being judged. This is a journey for any instructor. Neither of us began our teaching journeys as facilitators of learning, and we don't expect most of our readers to claim success in this regard either.

Unless formally trained in both theory and practice, for most teachers, developing a student-centered classroom is an exercise in trial and error. Early attempts may include cooperative groups, but just because we have students work together doesn't automatically mean this fosters learning. Groups may cooperate in some cases, but in other cases, one person may put forth all the effort and let the others copy the work so that all are relatively satisfied. This is not collaboration.

If you've ever attended a professional development session on student-centered learning, you have likely experienced talks about the difference between collaborative and cooperative groups. For those needing a refresher, collaboration is working together to accomplish an outcome or result. Cooperation is the process of working together, which is a necessary element of collaboration. For quality learning to take place, it is important that all parties contribute (cooperate) and understand all the pieces of the as they work toward the final outcome (collaborate).

Since most students are not conditioned to work in collaborative groups, it is essential to train them. Yes, you read that correctly—train them! Regardless of the age of the students, for a classroom to operate successfully, a teacher must define rules, norms, procedures and expectations. In other words, in order to develop a successful collaborative environment, a teacher must be able to effectively manage his or her classroom.

We suggest dedicating at least three days at the start of the school year or academic term, and even up to an entire week, training your students and conditioning their behaviors. Once a productive classroom environment has been established and students understand what is expected of them, it is then possible to focus on addressing content. This productive classroom behavior must be consistently maintained and will need to be revisited and reinforced throughout the entire year. It may take your classes months of redirecting, gentle reminders, and encouragement before they can be very productive.

> *A productive classroom environment should offer a climate of safety, where risk-taking is encouraged, trust and respect are fostered, there is open authentic conversation, and positive interaction is the norm. Classrooms should help foster intrinsic learning for students. Teachers will guide and support the students in positive behaviors.*

# THE FOUNDATION FOR MEANINGFUL CHANGE

## Kris's Classroom

The Center for Research on Learning and Teaching at the University of Michigan, on its website, www.crlt.umich.edu, states: "The first days of class are important in setting the tone for what is to come." As a teacher, I knew that I not only wanted to create a safe environment to help students overcome their fears and anxieties about mathematics, but I also wanted to create collaborative groups that fostered learning.

At the beginning of every school year, I started with icebreaker activities. I had several goals in mind, which I reinforced with my students throughout the year:

1. I wanted all students to talk to each other—not just to me.
2. I wanted each student to present to the group to establish that this was the expectation and not the exception in our classroom.
3. I wanted them to have fun. Math *is* fun! I wanted students to walk out of the first day of class saying, "Wow, was that math class? I think I'm going to like this class this year!"

With these goals in mind, I planned out my first three days of collaboration. By the end of the first week, I wanted my students to begin to identify with our activities and the work they would be doing all year.

## A Typical First Day

On the first day of class, I had students divide into groups in the four corners of the room in accordance with their birthdays—January to March in one corner, April to June in the second corner, and so on. In each of the corners, they were asked to introduce themselves to one another and share one thing they had done over the summer.

After a couple minutes, I asked them to pick a spokesperson who would introduce the whole group to the class. This first activity accomplished my first goal (above) and also helped me to determine four of my leaders for the year. (The spokespeople who were brave enough to speak to the class would most likely be our leaders.)

Once every group introduced itself to the class, I gave the spokesperson a king-sized candy bar as a thank-you for his or her bravery. This gesture was also to provide an incentive for other students to consider speaking up next time.

Speaking in front of peers is sometimes scary for students. I am not above bribing students for good behavior, and showing the class early on that I offered positive reinforcement from time to time only helped to build our student/teacher engagement. If you can't offer sweets, consider pencils, free homework passes, and similar rewards. Make sure that these are immediate rewards and not long-term incentives.

## The Foundation for Meaningful Change

Over the years, I noticed that students are more willing to volunteer in the beginning when this kind of behavior is rewarded. I found that I was able to transition students from extrinsic rewards to intrinsic rewards as engagement improved and students achieved successes along the way. Motivation requires making a connection to the heart, and motivating students is mission-critical if you want them to find joy in their work.

After the four-corners activity, I gave each student a colored index card. I then asked them to find someone with the same colored card and sit down and interview his or her new partner. I called this "The Business Card Activity," which allowed students to communicate with each other in a non-threatening, non-mathematical environment. (Instructions for this activity can be found in Addendum C, p 135). Follow the first part of those instructions. Be sure to collect the cards at the end of class so that you have them ready for the second day. This would also be a great way to take attendance for the day.)

On the second day, I had each student stand up with his or her business card partner and introduce this partner to the whole class. (I didn't tell them about this part of the task on day one, because I wanted them to concentrate on getting to know their partners, not worry about giving a presentation.) Each presented his/her partner's name, symbol, and one detail they had talked about during the interview.

At the end of the introductions, I had them high-five their partners and write their phone numbers and email addresses on the back of their business cards. Each pair became homework buddies for the year.

> If you are worried about privacy, consider modifying this to meet the needs of your classroom.

Day three proved to be the most important. I started the training for the collaboration that would take place on most school days by asking the students to create a chart sharing what they liked about group work and what they disliked.

I gave them one minute to think about and record their own thoughts. I then asked them to share with their partners for one minute. As groups shared, we recorded their thoughts on poster paper.

| Like | Didn't Like |
|---|---|
|  |  |

After a brief discussion, I gave the class the definition of collaboration, and we talked about how it differs from the "group work" and "cooperative groups" they may have

## The Foundation for Meaningful Change

experienced in the past. I then asked them to work together in groups of four to list some norms that are needed in order for collaboration to occur successfully. I asked each group to present one norm.

At the end of the class, I summarized the expectations for learning in our math classroom and referred to our classroom activities. In a nutshell, I told them there will be times when they'll learn individually and times they'll learn together collaboratively—but most important, I told them they will have fun learning.

I also explained to the class that while everyone learns at different rates, we will work together to provide an environment that allows each and every person to learn. My role as the teacher was to facilitate learning, and their role was to work hard because learning requires hard work and a lot of mistakes.

This was only our third day of class, and their training had just begun!

*****

Educators, this book will hopefully help you solidify and reaffirm understandings you've already learned, as well as add strategies to your tool box. Our purpose in writing these chapters is to give you practical examples you can use in your classrooms, but remember that every strategy will not be a perfect fit for every situation. As a professional, you must think about how a student-centered approach can be successfully implemented in your classroom.

What's always worked for me is to align my teaching to student learning by making learning interactive and all about the students. Most of the strategies I used on my first three days I learned in my first Carnegie Learning training, which I modified over the years to make them useful to my students' needs and our classroom. If I were to return to the classroom today, I would also consider incorporating some of the strategies discussed in Jo Boaler's book, *Mathematical Mindsets*—specifically her Growth Mindset strategies, which really work in the day-to-day classroom environment.

# THE SYSTEM

*This section was written with education leadership in mind—district administration, principals, and even those at the federal and state levels of our education system. As a parent, teacher, or student, you are welcome to read on or proceed to the next chapter about how we develop math learners.*

### RECOGNIZE

In a thinly veiled attempt to harness my own inner Jack Nicholson as Colonel Jessup in *A Few Good Men*, I ask you: You want the truth?

Can you handle the truth?

The truth is that our educational system is not designed to promote or develop numerical literacy. Most elementary teachers are passionate about and have a formal education in literacy. They are required to take classes on teaching reading and writing. Most enjoy reading and writing and are excited about teaching their students to learn to read and write.

Then, there is math. While teaching math in elementary grades is a requirement, teachers often do not receive anywhere near the same type of formal training for math as they do for literacy—much less ongoing professional development. If you ask elementary teachers to self-report on their academic prowess, and the grades that they received in math when *they* were students, it has been our experience that the majority of responses will fall in the B/C range.

Conversely, we have our intermediate and secondary teachers. By intermediate school, and most certainly by secondary school, math courses are taught by math specialists. These are teachers, who, as students, liked math—a lot. They were A students, and if not A students, certainly B students. They were able to memorize formulas, solve equations, and conquer the math universe. At some point, the idea came to them, "Hey, I am good at math. Very good actually. I should become a math teacher!"

**UNDERSTAND**

Regardless of grade level, we would argue that in order to teach, you should first be an expert. However, our current system does not see it this way—especially with respect to instruction at elementary-grade levels. Instead, elementary teachers are required to possess a basic competency and a willingness to present math information. For many elementary teachers, what this equates to is, "Give me the resource and I will tell students what it says."

The better teachers understand the math concepts they are being asked to present, the better they are at actually teaching the concepts. Interestingly, our system doesn't require them to be able to "teach" math—it only requires them to present math concepts.

## The System

Although it sounds great to have a high-achieving lover of math serving as a specialist for middle and high school classes, this can also be a curse. Why? Because not all students are going to walk into that teacher's class with the same love of math or with the necessary foundation to hit the ground running when they walk into Mr. or Ms. Math's classroom.

For teachers, we have observed that fewer than 20 percent of the students they are attempting to teach are "their people," fellow lovers of math. With the majority of their students falling somewhere below their desired learner range, intermediate and high school math teachers often find it difficult to relate to students who struggle with or are disinterested in math.

Some teachers lack the ability or desire to help struggling students develop alternative strategies that might be better suited for such students. They think, "This is how *I* learned math, so this is the way my *students* need to learn math." Some teachers hold onto unproductive beliefs that focus their time, attention, and affection on "their people"— the 1 out of 5 students out there who love math.

### Manage
### Kris's Classroom

Overall, teachers have very different approaches when it comes to professional development due to their vast range of experiences. Some teachers lack content expertise, some lack the ability to engage students, some struggle with the "how" of mathematics, and some have a hard time with the "why." Professional development for teachers and instructional leaders is a key component for schools to increase mathematics understanding in students.

What do you remember about your mathematics education experience? Most of us experienced mathematics in a more traditional fashion, like I did in Mr. Frank's classroom, than what I experienced in Ms. Patterson's class. In James Stigler and James Hiebert's book, *The Teaching Gap*, the authors point to research that shows the majority of U.S. classrooms follow the following structure: warm-up, check homework, demonstrate a procedure with definetions, students practice the procedure, homework.

The more disturbing part of this research is that 80 percent of the mathematics concepts are only stated, not developed, and 95 percent of seatwork is engaged in practicing routine procedures.

How do we expect our classrooms to change if our professional development stays the same? How often do teachers sit and get in their in-service days? Overall, our approach to supporting teachers and leaders must change if we expect our students to change.

Most of us experienced mathematics in a teacher-centered classroom. How do we move toward a more participant-centered professional development plan so that teachers can experience what we want them to do in classrooms with their students?

## The System

Adult learning is much more difficult than K-12 student learning. As adults, we have blind spots and have been practicing some bad habits for many years. I have also discovered that K-5, 6-8 and 9-12 teachers have very different needs.

It is critical that we differentiate professional learning for our teachers. Not only do these grade bands have different needs, but there will also be individual needs within these grade bands. It is important that as instructional leaders, we consult experts and incorporate different types of professional development for teachers.

Elementary teachers make up one of my favorite groups of teachers to do mathematics professional development with because their "light bulb" moments are events we celebrate enthusiastically. When I see an elementary teacher's fear of math turn into excitement and confidence, it's awesome! I love when they shout, "I got it! Let me teach you my method."

Most elementary teachers do *not* love mathematics. I've had many tell me they went into elementary education because they hated math and loved reading. I have listened to sad stories about bad experiences with teachers when they were in school and heard too many of them say, "I don't have the math gene."

When I work with K-5 teachers in workshops, I have to keep in mind that math anxiety is sometimes at its highest among this group. (This is especially true for grades K-2.)

These teachers are not typically content experts; they are generalists. The good news is that the great teaching strategies they are using to teach science and other subjects can also be used for math.

An algorithm is an unambiguous specification on how to solve a problem. For instance, when adding fractions, you first find a common denominator (bottom number) to create a new equivalent fraction, then add the fraction numerators (upper numbers). Lastly, the new numerator is placed over the common denominator.

Elementary teachers will typically be open to learning multiple methods and discussing their reasoning because they aren't sure of the algorithm. If they do know the algorithm, they are the first to say, "I have no idea why it works. That is just the way I was taught to do it."

We always have some good laughs when I ask teachers to explain. Their response is, "My 3rd grade teacher told me so."

Elementary teachers are also typically fearful of letting students venture too far from the paths with which the teachers are familiar. This is because they aren't sure how to answer questions, and sometimes they aren't sure if the students are correct. I have been in many elementary classrooms where the teacher tells students they are wrong, but never asks for their reasoning.

When I coach teachers after class, I always ask, "How do you know the student was incorrect?" If the teacher is unsure of the mathematics and hasn't heard the conversation in the group or the explanation by the student, the teacher cannot judge correctness solely based on the answer.

It is essential in professional development and professional learning group time that teachers explore the mathematics within the lessons they will teach for their own understanding. I sometimes have teachers do the math task before our meeting, and occasionally we do it together. The goal is to talk about the mathematics and our multiple approaches to the answers.

It is also important to have teachers discuss how they will facilitate tasks without lowering the cognitive thinking or rigor of the task. To support this continued learning, I find it important to model or co-facilitate with teachers in their classrooms.

The high school group is typically the opposite of the elementary group. As a high school teacher, I was definitely one of the difficult know-it-alls. When I work with high school teachers, I always take time to affirm that they are content experts. High school teachers know their mathematics, especially the algorithms. The blind spot is that while we have memorized the algorithms, processes, and typically know the language, we are missing the conceptual understanding and the connections to elementary mathematics. We also usually lack the student-centered focus that elementary and middle school teachers embrace.

When doing workshops with high school teachers, I find it important to model a student-centered classroom without treating participants like children. Give participants time to think on their own and talk with a partner or group, then present the information. Ask participants to explain their reasoning, and encourage others to add to the conversation.

It is also important to randomly select teachers to present their ideas and circulate while they are sharing in their groups. Many teachers will talk in the smaller groups but are not confident enough to share with the room full of math experts. With these teachers, I explicitly have them chart the "teacher moves" I have been modeling and talk about why each is important.

In a recent conference, I facilitated a "teacher moves" debrief in a workshop. One of the participants raised a common misconception among teachers. This teacher said, "Kris, this is all great discussion about teacher moves, but let's be real—we are adults, so we are going to behave and do what we are supposed to do. Kids don't behave the same way." Another teacher spoke up and said that teachers are the worst students.

One teacher had been off-task during the gallery walk (viewing of students' work). When I asked the groups to list the teacher moves, this teacher shared with her group that she had been distracted and the facilitator had redirected her to make sure she was working with her group.

# The System

When I asked the teacher to share how the facilitator redirected her, she said, "Exactly how I do it in class. She asked me a question, then told me my group would love to share in my discussions. I received a compliment and a shove at the same time. It was perfect."

Some high school teachers also need a reminder that math is FUN! This is the time to allow students to explore these really cool functions, how functions behave, what their characteristics are, and how they connect with one another.

I once had a high school teacher tell me that manipulatives are for elementary grades only. I disagree wholeheartedly. I used algebra tiles, pan balances, popcorn, clay, measuring tools, and many online tools to help support student learning. I've also had students act out the different functions. For example, we'd go outside and measure shadows and physical height, then use their knowledge of geometry to find the height of the flagpole.

In high school, teachers know their mathematics content, but they usually need some ideas on how to make math engaging for the learners. When I work with high school teachers in their classrooms, I rarely model unless they ask. When I do model, it is usually with instructional strategies, classroom questioning, or the facilitation of collaborative learning.

High school teachers sometimes struggle with planning. They are asked to wear many hats, and sometimes these extra hats take over their focus on planning lessons. Since these teachers usually can "wing it," they do little planning, which results in missed opportunities for learning. They also need to take time to plan the whole unit (or several chapters) to understand the flow of the mathematics and the connections.

If you're an instructional leader, do not allow teachers to spend five or six weeks reviewing at the beginning of the year. Start on the standards and infuse the "review" as needed throughout the chapters and lessons.

Let us now think about middle-school teachers. I firmly believe there is a special place in heaven for these teachers! Teaching middle-school students, whose hormones are starting to awaken, takes an hourly dose of patience and perseverance.

My son is finally moving out of middle school into high school this year, and I'm so thankful. These years are the toughest on parents and teachers. As instructional leaders, keep in mind that teachers at the middle-school level also have very different needs than those teaching elementary and high school classes. Some come to the table as generalists and are certified K-6, while others are 7-12 certified.

As an instructional leader, you need to know your teachers' needs. Do they have a lack of content, confidence, engagement strategies, or rigor within the tasks they have the students explore?

# The System

> **Rigor:** The right mix of conceptual understanding, procedural skills and fluency, and applications to address the standards appropriately for a particular grade.

At all levels, teachers need to have a great set of resources with which they can facilitate their math classes. However, at the middle-school level, resources must include an overview of the mathematics, prior mathematics, connections, and strategies for facilitation. If they don't, then professional development and coaching will more frequently be a necessity. These teachers will have such diverse needs that true differentiation will be needed.

## I'm Just Not a Math Person!

Most schools see a dip in achievement as students come into middle school. Not only are the students changing, but the teachers have very different needs. Consider providing workshops and, more importantly, side-by-side coaching for these teachers. I know this can be a big cost, but the reality is that if you want to support teachers, you have to invest in their learning as well.

When I work with teachers to plan lessons and then model, co-facilitate, and observe, I find the biggest impact on their own learning. Workshops must be customized and cannot be a one-size-fits-all.

Math teachers typically feel left out in professional development with the emphasis on literacy in our schools. As an instructional leader, you have to consider the different levels within your mathematics department. I highly recommend bringing in an expert to facilitate professional development and continue that support throughout the year.

Some good news at the middle-school level is that it is a great place for building collaboration and visiting classrooms. As I mentioned above, elementary teachers usually embrace more engagement strategies and the collaborative classroom, while eighth-grade teachers probably have a solid content understanding. When you merge those two, you have your model teacher. I find that teachers enjoy watching one another teach, and they learn a lot from these experiences.

## THE SYSTEM

Some veteran teachers have been practicing direct instruction for so long that changing their practices will take longer than it will take a new teacher to learn fresh out of college. New teachers usually struggle most with classroom management, which will sometimes cause them to turn to more direct instructions and a teacher-centered math classroom to keep control.

A huge factor in teacher development is *mindset*. It will be much more difficult to improve the practices of a teacher—novice or veteran—who has a fixed mindset. On the other hand, teachers with growth mindsets will improve their practice more rapidly.

Ultimately, we, as teachers and instructional leaders, need to focus on lifelong learning. We are teaching students today for jobs that have not yet been invented. No one my age (I grew up in the 1980s and '90s) thought they would work for Google, for example, because we didn't even know much about the Internet.

Many of you reading this book didn't grow up with a cell phone. If someone had told you that you would carry around a mini-computer in your hand, and feel lost without it, you would have laughed.

The world is changing rapidly, and education must keep up. We need to support teachers in preparing students to become critical thinkers, to persevere, and to love learning so they can be successful in the future.

### Questions for Reflection
### Administrators

- How is professional development (PD) delivered in your school/district?
- How will you model the teaching "moves" you want to see in classrooms during your PD time?
- How can you provide math expert support for your teachers to help move them from a teacher-centered classroom to a student-centered classroom?

# How Do We Develop Math Learners?

## Recognize

For starters, we need to recognize that our current system is not working—at least not as it should or could be for many of our students. Specifically, only 32 percent of U.S. students from the 2011 graduating class achieved math proficiency, compared with 56 percent of students in Finland, 58 percent in South Korea, and 75 percent of students in Shanghai, according to Paul Peterson of EducationNext (www.educationnext.org). The United States placed 32nd in math among the 65 nations of the world that participated in PISA, the math test administered by the Organization for Economic Cooperation and Development (OECD).

## Understand

### The Teacher Conundrum

We need to understand why our current system is not working if we have any chance of fixing it. For starters, our classrooms more closely resemble a factory assembly line than a place or operation that is custom-order. The assumption that all kids have the same parts, can be put together the same way, programmed with the same information, and sent out into the world as fully functioning professionals is simply preposterous! Every student—and every person, for that matter—is unique.

Each student comes into the classroom with different needs, different dreams, different academic and personal experiences, and different ways that he or she learns best. Kids are not machines, and the fact that our classrooms, more often than not, are designed to present information, have students memorize the information, and recall it well enough to pass tests, is completely absurd. Yet, for far too many students, this is precisely the environment they find themselves in each and every day.

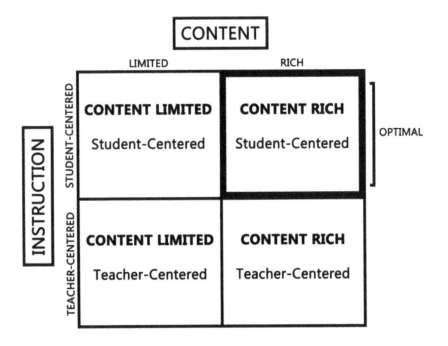

## How Do We Develop Math Learners?

In our experience, most elementary teachers fall into the top left quadrant, while many middle-school and most high-school teachers fall into the bottom right quadrant.

Conversely, there are teachers who do more than present facts, figures, equations, algorithms, and the like. There are teachers who break free from a traditional "sit and get" environment and look at each and every student in their classrooms as individuals.

These teachers, these wonderful teachers, understand that true learning and differentiation requires them to be classroom facilitators—not presenters—in order to help students develop their own rich, conceptual understanding of the information they are expected to learn.

So what is it that makes these teachers special? What is required to facilitate learning in the classroom?

## Manage

We contend that there are four instructional pieces necessary to reach students on a deep, personal, and meaningful level.

1. **Teach WHO:** Get to know your students.
2. **Teach WHY:** Understand each student's goals, dreams, and sense of self.
3. **Teach WHAT:** Facilitate learning of the subject material.
4. **Teach HOW:** Create processes, procedures, and computations based on the needs of your classroom.

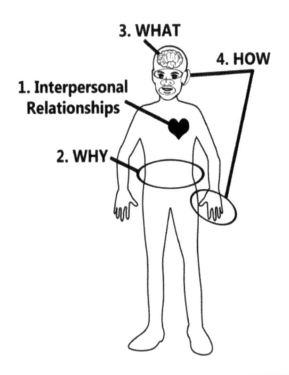

# How Do We Develop Math Learners?

## 1. Teach WHO: Teaching to the Heart

Before students learn a thing, they must engage. Students who are entirely checked out mentally, physically, or both, are incapable of learning anything.

The best way to get students to engage is by building authentic, positive connections. When individual students feel like teachers have their best interest in mind, and that these teachers are committed to their success, students will engage. The single-most critical thing to keep in mind is the importance of trust.

As a teacher, you must earn your students' trust, and here is the bottom line: if you are not interested in making personal or meaningful connections with your students, then get out of the classroom. This may sound harsh, but positive teacher/student relationships are essential in helping students reach their potential. There are plenty of other jobs that do not require us to get to know others on a personal level. Teaching is not one of those professions.

## 2. Teach WHY: Teaching to the Core

What do your students want to be when they grow up? What are their goals and dreams? If you have 25 students in your class, it is more likely than not that they will eventually work in 25 different jobs for 25 different companies. Some will pursue a four-year college education, while others will not.

As you teach concepts—any concepts—it is important that you, as a teacher, help students make personal connections to careers and activities that are of interest to them. This macro, big-picture view helps students develop their personal understanding and appreciation for what is being taught.

### 3. Teach WHAT: Teaching to the Brain

For most teachers, this tends to fall within their comfort zone. This is the content itself. It can come directly from a textbook, from supplementary handouts, or from just about anything that can be considered a resource. This is the micro view and procedural lessons that help build a student's ability to use what is learned for practical purposes.

### 4. Teach HOW: Teaching Multiple Strategies and to Multiple Learning Modalities

It is our assertion that this area of HOW is precisely where we, collectively as an education system, struggle the most. We need to build into our curricula teaching students different strategies for HOW to learn—and getting them to recognize how they learn best.

It is not enough to teach students *what* to learn, or even why those topics are important. Learning preferences are often talked about broadly in education as "learning modalities" or "multiple intelligences," but seldom, if ever, are they considered part of daily instruction. We need to be aware of learning strategies and actively engage in our own learning. But to do that, we need to teach students strategies to help them become successful learners.

Because all students are different, and have different strengths and weaknesses when it comes to the use of different learning modalities, it is important that we incorporate activities that use different, and even multiple, strategies whenever possible throughout our instruction in order to best connect with a wide range of learners.

We can all learn, but we all learn differently. As a system, due to limited time, and in some cases, limited teacher training, this could be our greatest challenge yet.

Students spend the majority of their day being told what to think—and often, why what they are being taught is important. Rarely, however, do we spend time helping

To engage students and make math meaningful, we need to teach math consistently in multi-sensorial ways. We need to speak to the heart to stimulate the mind.

We have to appreciate the whole learner and not assume that all students can be programmed to learn in outdated and traditional manners.

students identify their learning strengths or teach them strategies to help them learn best.

If we have any hope of helping students develop a deep conceptual understanding of information, and the ability to retain it for productive use at a later time, then we must encourage the development of learning environments that promote healthy interpersonal relationships with teachers. We must teach students the macro view (WHY) and the micro view (WHAT) and be certain that we appreciate how every learner is different. We must help students learn to develop strategies that help them learn best.

The HOW transcends math. Learning how we learn best is something that can be applied throughout life, regardless of content or subject matter. Learning how to learn would also demonstrate to students an appreciation that we are all different and unique, which would help develop a greater sense of self as learners. This is the total student approach.

## Student-Centered vs. Teacher-Centered Teaching

As teachers, we need to believe at our core that learning is about the student and not about us. We each have our own comfort zones in the content, and we also have our comfort zones in the way we teach. Most of you reading this book were not taught in a student-centered classroom, and the majority were likely taught in teacher-centered classrooms and conformed to that way of learning or memorizing to be successful.

## How Do We Develop Math Learners?

A true student-centered classroom focuses on how students and people learn, not on how a teacher has been taught. This will look different than the traditional (and passive) sit-listen-and-take-notes participation for students. A student-centered classroom will have students talking to each other and presenting their ideas to the whole classroom. In this setting, teachers will facilitate the conversations and ask questions to probe and create discussion.

In order for a student-centered classroom to be successful, teachers must know their content and facilitate the learning—and the students must be actively participating. A student-centered approach encourages authentic learning, and authentic learning is what resonates with students and helps them succeed beyond a class, beyond a subject, and beyond school.

### Kris's Classroom

In reaching students and helping them understand WHY, and teaching them HOW to think—it is not about teaching them WHAT to think. Our current system—just like most of our own educational experiences—focuses too much on WHAT and not enough of WHY.

As teachers and instructional leaders, we sometimes believe that since teachers have a degree or certification in mathematics, they deeply understand mathematics.

For me, that was not true. I came out of high school with awesome grades in mathematics because I could memorize. I got good grades in college, too, because I was good at memorizing and then regurgitating information on tests. It wasn't until I took my elementary mathematics methods course, sophomore year in college, that I realized I had no idea how math worked conceptually. I realized I knew the algorithm. I had memorized the facts—but conceptually, I had some work to do.

When I started teaching high school, students would ask me WHY? "Why does that work? How does that work? Where am I ever going to use this math? Why do I need to learn this?"

I loved math in school. I loved manipulating equations and proving theorems, but WHY wasn't really anything I had thought about much.

I taught at a vocational school for my first 10 years in education, so I really wanted my students to explore how mathematics related to the career paths they had chosen.

Once, my class was working on a geometry unit, exploring volume and surface area. I decided to ask the horticulture teacher to help me create a unit for his students that was directly related to this concept. I was excited to engage the students in mathematical discovery and connect their understanding of horticulture to the mathematics they would use in their careers. I was also pleased to see the higher level of engagement with the mathematics concepts when I connected the mathematics to something they were interested in learning.

Developing conceptual understanding and procedural fluency are important in a well-balanced mathematics classroom. We have learned from research that concepts should be explored and investigated so that students can derive their own formulas, algorithms, and reasoning. *Principles to Actions, Ensuring Mathematical Success for All* (the National Council of Teachers of Mathematics's landmark publication) mentions productive beliefs and unproductive beliefs. Many educators have practiced the unproductive belief for years. In fact, our society also looks at mathematics as definitions, formulas, and rules to follow.

As educators, we need to engage students in mathematics. We need to create environments in which students want to learn. One of the ways we can make math FUN is by engaging students in tasks that are in context and allowing them to explore and grapple with the mathematics.

When students worked on real-world tasks in my classroom, I found that they took time to make sense of the math, and when asked to transfer this knowledge to a new situation, they were able to do so. As I mentioned previously, this perseverance took time. As the teacher, I wanted the students to make sense of the mathematics so that at a later time (the next day, next week, and next year), they could successfully use the mathematical ideas and apply them to new contexts.

For teachers, this is not easy to facilitate and takes some practice. I often reminded myself that I wanted the students to enjoy mathematics, not just regurgitate the facts and algorithms.

Over the past 18 years, I have been facilitating professional development sessions for elementary, middle, and high school teachers. Some of the most rewarding workshops are when teachers learn math conceptually for the first time. The teachers in my workshops love to find the correct answer using algorithms—and sweat a little when I ask them to draw a diagram or model to prove that their algorithm works.

My favorite question to participants is, "How can you prove your answer is correct?" Some say, "I just know it is, because that is the correct procedure." My response is then, "How did you know that was the correct procedure? How can you prove that that procedure will always work in this situation?"

I strive to get participants to look beyond the correct answer so that they can consider the concept as a whole instead of just getting the answer. When they figure out how the procedure works, it is their light-bulb moment. They are so excited and even more proud when a few days later, they can make the mathematical connections because they now understand the concept, not just the procedure.

After we explore the concepts in these workshops, we take time to reflect on the instructional "moves" the teachers have experienced. *Task selection* is number one. An instructor must pick a higher-level thinking task to allow students to explore the concept. *Facilitation* is second in importance. I ask teachers what I, as the instructor, did to support their learning.

Comments included:

- "You didn't answer our questions; you asked us another question to guide us."
- "You listened to our conversations during small group time, and during presentations, you asked one of us to share our ideas to help us make connections."
- "You had us present our learning, and you asked our peers to validate correctness and that our reasoning made sense."
- "You made math safe and FUN by encouraging us just enough to keep us going."
- "You didn't really care if we had a final answer yet. You wanted us to explain what we were doing and why. The correct answers were derived as a whole group sometimes."

As educators, we need to understand that we can't teach conceptually by telling students what to do. The stand-and-deliver style of instruction will not allow students/participants to explore the concepts. Therefore, what we produce with this method of instruction is mostly algorithmic or procedural fluency. For conceptual understanding to happen, teachers must allow students to explore the concept while facilitating the learning—or as some say, be a "guide on the side."

As the facilitator, I never want students to get frustrated to the point that they quit—but I do want them to struggle and persevere. I want them to make sense of the mathematics and not just focus on finding an answer.

The past of couple of years with my son in middle school have been an interesting journey. Although I could write another book about his overall behavior, I will stay focused on his mathematical thinking. In middle school, he permitted me to help him with his homework on occasion and he always started by saying, "Mom, I don't want this to take two hours. Just tell me how to do it."

I struggled with "just telling him" how to do it, because this skips the opportunity to understand the concepts. One time, I decided to help him connect to his learning from the classroom instead of re-teaching my way, conceptually. I asked him where his notes were from class. We started going back through his notes on factoring, and I pointed out that the problem we were doing was exactly like the example she did in class.

He said, "I don't have time to listen to her in class because I have to take all these notes to get my notebook grade."

I was floored he admitted that. He was taking notes, not for the purpose of referring back to them to understand, but because they gave him a grade.

## How Do We Develop Math Learners?

While teaching, I did not grade notes, but I admittedly did grade some irrelevant things that had nothing to do with mastering or understanding the content. I did not fault my son's teacher, because she probably didn't understand that he was not using these notes purposefully.

There are many books out there about grading, and my philosophy has always been this: allow students to redo and retake every assignment or test/quiz until they have mastered it. I want students to master the concepts and skills. Students get very discouraged when they see that they did not do well on a test. Some students will act like they don't care, but they do.

As much as I encouraged students to show their work and talk to the class about what they were thinking, I also had them redo work they did not quite understand. I allowed students to retest several times, if needed—but they had to come and receive some tutoring after the second try if they wanted to redo a test for the third time. My goal was to build up their confidence with regard to the fact that they were still learning. Because I care about mastery of the concepts, if they didn't understand these concepts on the Friday I gave the test, we would continue to work together until they mastered the concepts—even if that took another week.

As you reflect on your own teaching practices, ask yourself a few of questions:

- Are my beliefs about teaching and learning productive or unproductive beliefs?
- Am I making math FUN for students? If not, how can I make math FUN?
- Are my students learning the concepts or just following procedures and learning algorithms? How can I facilitate conceptual understanding of the mathematics?

If you struggle with good answers to these questions, you are not alone. It took me three years of teaching before I started seriously asking myself these questions. I am still working daily to learn more and practice new ways to help teachers and students explore mathematical concepts.

By reading this book, you have already started to explore a deeper understanding of aligning your teaching to student learning. I think my favorite way to learn is to do math with other teachers and students. I love hearing what they are thinking and why they solve tasks the way they do. I like to listen to their reasoning, even if it isn't correct. There is always more to be learned from an incorrect answer than from a correct one.

If you don't have others with whom you can work on mathematical tasks, then join an online group or follow other math teachers on social media. The key is to work on math tasks yourself so that you can explore the mathematics.

# How Do We Develop Math Learners?

## Productive Struggle

As students explore mathematical concepts and ideas, there will be a natural struggle with certain areas within their learning process. Students should be allowed to struggle individually and with their peers as they wrestle with mathematical concepts. In other words, teachers must allow them to struggle "productively."

Productive struggle is critical to supporting student thinking and learning in mathematics. As teachers, we have a natural desire to "help" children to "get it," and we want to "save them." In my math classroom, the more I allowed students to productively struggle, the more they were able to apply their learning to other problem-solving tasks later. In business, the best leaders fight the urge to take over and save people while they struggle.

The *Task Analysis Guide* states that, "Doing Mathematical Tasks" requires considerable cognitive effort and may involve some level of anxiety for the students due to the unpredictable nature of the solution process required." (Mary Kay Stein, Margaret Smith, Marjorie A. Hemingson, Edward A. Silver: *Implementing Standards-Based Instruction, A Casebook for Professional Development*, 2000). Tasks classified as "Doing the Mathematics" promote problem solving where frustration potentially will occur.

Our role as teachers is to support students in persevering without taking over their learning. On the other end of the spectrum, we also want to support their struggle so that they do persevere and do not give up. This can be done in a variety of ways.

First, there should be clear expectations. Teachers must make sure that students understand their responsibility in the learning process—namely that they have an obligation to make sense of the mathematics.

General expectations encourage students to make sense of the mathematics individually, in small groups, and as a whole class. In my classroom, the students knew they needed to start by reading the scenario or task, try to make some sense of what was asked of them, and discuss possible strategies and solutions with their group members.

During my first couple of years teaching, no sooner had I assigned a task than hands would shoot up from students who didn't understand. To encourage them to think before asking, I implemented a few group strategies: red/green cards, "Ask Three Before Me," and writing down a question instead of saying, "I don't understand."

- **Red/Green Cards**: Students are separated into groups, and each student group is given a red card and a green card. The green card stays on top until the student gets stuck, at which point the red card is placed on top. When all group members have red cards on top, it's time to offer help.

- **"Ask Three Before Me"**: When students think they need help, they are to ask three other people for help (usually their group members) before they raise their hands.

- **Question Writing:** Sentence stems and questions are provided for students to ask within their groups.

When you set clear expectations for students, they will usually meet them—if not exceed them. While it might take a month or two, students will eventually explain and discuss math problems with little or no prompting. I knew I had set up clear expectations when one day I overheard one of my students tell a new student, "Don't raise your hand and ask her until you have talked to your group first. She always asks us, 'What did your group discuss or think?' You'd better be ready to give her a real answer, not just 'I don't know.'"

Second, students need to understand that being able to explain and justify their reasoning is as important as—if not more important than—having the correct answer. Teachers should provide students with access to manipulatives—or tools—that can support the students' thinking processes. These teaching tools engage students visually and tactilely, such as include coins, blocks, counters, tiles, technology, graphs, and tables. Students should be encouraged to use manipulatives and sketch or draw as they work in order to help them find solutions.

National and state assessments have changed in that students can select multiple correct solutions and not always one correct answer. Teachers can help students focus on their reasoning and have several correct answers.

They can also give students the numeric answer and ask students to explain how they would arrive at that answer.

Third, students must be able to effectively communicate with their group members to discuss and defend their thinking. Effective communication among student group members about their own thinking processes helps the group move forward and make progress on a task. Within their groups, students should be able to explain their reasoning so that other members can add to the strategies and discuss correctness, and students must listen and be willing to critique the reasoning of others. Using this model, teachers provide a safe environment for students to ask questions that probe deeper into their thought processes.

Teachers sometimes take over the learning process by sharing their ideas and solutions before allowing students time to grapple with their own understandings. It is important that teachers ask students to explain their reasoning and are careful to focus on answers—not make assumptions based on what they want or expect to hear. I used to use the phrase, "Tell me more about..." to have students rethink or restate their own thinking. When a group got stuck, I asked them to explain what they had discussed so far before jumping in and "helping" them.

The toughest part of productive struggle is the "productive" piece, which is best done when students are given the opportunity to discuss and validate strategies and solutions. This will look different for each student and classroom, and what worked for me might not work in other classrooms. Students are unique individuals, so each teacher should have a toolbox with as many tools as possible to support them individually and as a group.

Listening, rephrasing, and questioning student thinking are key to making the struggle productive. Knowing when students are ready to quit and how one question or redirection will help them to persevere is why teaching is a craft. As teachers, we should continue to learn and find new strategies to help support students so they love learning—especially math!

### *The Benefits of Productive Struggle*

If you are a teacher or have a background in education, you have likely heard the term "instructional range."

If students are routinely given concepts and tasks that are too easy, their minds and hearts will not be sufficiently stimulated for them to remain engaged and on task. If concepts and tasks are too difficult, and they do not have the basic foundation or prerequisites to make sense of new information and tasks presented to them, they will likely disengage—and their achievement gap will widen as time goes on.

> A student's instructional range is essentially a sweet spot for learning in which concepts students are being asked to make sense of, and tasks that students are being asked to successfully complete, are neither too difficult nor too easy for them to undertake.

>  Identifying and teaching within an appropriate instructional range for each student in a given classroom is essential. Teaching below or above a student's instructional range for too long can cause a student to disengage, and when students are physically and/or mentally disengaged, then we fail to help them reach their potential and make the most of the time we have with them in our classroom.

When students are asked to only provide answers on a test, this represents *convergent thinking*, or thinking that produces a single correct solution. When students are provided tasks and assessments that generate a creative thinking process and they explore different ways to solve problems, this produces new ideas, patterns, and an extension of cognitive networks. This is called *divergent thinking*.

Assessing only convergent thinking ignores the development of higher order thinking, problem-solving, experi-

> Data-driven instruction is a cycle where students are assessed, teachers and students analyze the data, then create a plan of action to remediate, accelerate or address any needs. This cycle should continue frequently.
>
> With good data in hand, teachers can focus on dynamic grouping and facilitate learning that encourages a productive struggle to meet the needs of a wide range of learners in a given classroom.

menting, and synthesizing skills. If students have no meaningful reason to retain information over time, they will continue through a system in which they study for a test, take the test, and move on. When students discard information at a later point, meaning is not created or retained in long-term memory—which means future understanding or application is unlikely.

An important task for teachers is determining, on a consistent and ongoing basis, where students are in their learning and addressing their needs within a targeted instructional range. This is why data-driven instruction is so important and why capturing quality data is meaningful for teachers and students alike.

## CAN VS. CAN'T

At a time when universal screenings, assessments, and diagnostic tools are all the rage, it is easy to see why we hear so many teachers say things like, "My students can't...." The notion of "my kids can't... " is especially common in the world of special education.

While diagnostic tools are great for pointing out what students *can't* do, they are also extraordinarily helpful in helping us identify what students *can* do. Unfortunately, we are often so hung up on what our students *can't* do that we lose sight of what they *can* do.

Learning doesn't occur in a vacuum; it requires exposure and experience. The more we are exposed to, the more

> **Administrators**, you are probably reading this book in hopes of supporting your mathematics teachers. I would implore you to also consider that productive struggle is important in adult learning and teaching as well. Teachers also become frustrated and quit tasks when they are not supported.
>
> - At the elementary (K-6) level, this frustration might be more focused on understanding the mathematics themselves, because most teachers are generalists at this level.
> - At the middle and secondary level (7-12), this frustration might be focused more on teaching practices of the concepts.
> - Professional development is very important in supporting teachers, not only in workshops with mathematics learning but also side by side in their classrooms.
>
> As administrators, we can hire math coaches who specifically help with developing math teachers' pedagogical understandings along with their content development. These coaches should be focused on supporting teacher learning that directly impacts and improves student learning.

we experience, and the more deeply we form and understand important lessons, the more capable we become.

As teachers, we must move away from dwelling on what students cannot do. Conversely, we need to do a better job of embracing and understanding the things they can do and help build on that foundation. It is for this reason, among others, that personalized learning and differentiation are so immensely important.

Focusing on what a student can do is empowering to a child. The "can" moments and connections offer small wins, which positively reinforce engagement and effort, and in turn, help a student continue to progress. As new concepts are introduced and students begin to struggle, these moments will be positive capital investments—resulting in students not quitting or giving up as quickly.

Within every lesson, for every child, there is a "can." While there might very well be more can'ts than cans, we as teachers cannot dwell on the can'ts because when we do, we are setting the example for our kids. Every time we believe they can't, we are reinforcing their own belief that they can't as well.

In every lesson, we can identify cans and can'ts. Make an effort to move instruction toward the positive while maintaining awareness of limitations. Help students identify the things they can do, and work with them to build upon it. In time, students will increase cans, reduce can'ts, and have a much more positive outlook on their own abilities—including math!

## Closing Exercise

### *The Ideal Classroom*

As a teacher and parent, it is easy to get caught up in habit and routine. Teaching from the same textbook or set of resources for several years makes it even more likely that you'll get caught up in pacing and procedural tasks. The same goes for compliance measures.

Our education system is so heavily focused on compliance measures, and not focused enough on actual learning, that it tends to have a less-than-desirable impact on how teachers manage their time on a given day.

Take a step back. Put on your "learner" hat for a moment and think about your ideal classroom.

- When you envision the perfect classroom, what does it look like? What does it sound like? What does it feel like?
- At the end of each lesson, each day, each year—what reflections would you expect to have about your learning experience in your ideal classroom?
- How would you describe the teacher who is teaching in your ideal classroom?

Now, teachers, put your teacher hat back on. Be reflective and introspective, and even perhaps a little vulnerable, as you attempt to answer the following questions:

- How do your students describe their experience in your class?

- Is it consistent with your expectations of your ideal learning environment?
- What are some words or descriptors they use to define you as a teacher?
- What instructional strategies do you believe describe your approach to instruction? Why do you feel that way?
- What do you believe is the biggest barrier to achieving your ideal classroom?

Now, parents, take time to reflect:

- How can your words and actions support your child's learning experiences?
- How are you encouraging your child without trying to take over the teacher's responsibilities?

## Kris's Classroom

### *Final Thoughts for Educators*

We need to provide our students with learning opportunities—then get out of their way and let them productively struggle so they can become proficient in all eight of the Standards of Mathematical Practices (see Addendum).

One of my biggest challenges as a teacher was NOT taking over the learning for the students. I would see them struggling, and I'd want to help—but it's important to resist. Don't take their pencils and do the math for them. Don't be a "sage on the stage," and never say, "My students can't..." These are unproductive beliefs in education and bad practices.

If you really want to engage students, do the math yourself first. This should be the first step in your lesson plan. Do the math, and then design good questions to uncover misconceptions. Practice wait time during the facilitation, and make math FUN!

For us to engage students in mathematics, we must know the math ourselves and have a toolbox of great resources to offer them. Sometimes I'm baffled when I'm talking to school administrators who say, "I don't know why the majority of my teachers don't have a collaborative classroom." We then start exploring their resources and

---

**Parents, please don't try to teach your kids the math!**

Most likely, we didn't learn conceptually so we will typically go directly to the skills and sometimes cause more frustration and confusion.

I think of it this way: My son, Nate, just started learning to play the guitar. To help him, I have not picked up his guitar, nor have I started to learn to play the guitar myself. Instead, I have provided him a teacher who gives him lessons.

My job is now to ask questions like, "Have you practiced yet today?" or "What did you learn in your lesson?"

I will encourage him after listening to what was supposed to be *Thinking of You* by Ed Sheeran (but I didn't hear anything close to that). I will tell him to keep working hard and practicing!

**My job as a parent is to ask questions and provide encouragement!**

textbooks, and find that their resources do not support teachers in this shift to use a collaborative classroom.

If educators pick higher-level thinking tasks that are relevant to their students, then engaging students in mathematics is much easier. There are only a few textbook publishers that focus on engaging students in conceptually understanding the mathematics before moving students to fluency and process. Most focus more on fluency of algorithms and procedures along with correct answers. If students are conditioned to focus only on getting correct answers, they will disengage from the learning process after several failed attempts. If we want them to persevere, then we must embrace the exploration, encourage them to explain their methods, and look for multiple ways to solve tasks.

Administrators also need to model best practices and provide ample professional development for teachers to be successful. True professional development must be differentiated and designed to meet the teachers' needs. Teachers, like students, are tired of sitting through another boring day of professional development.

How do administrators provide professional learning with choice, outcomes, and follow-through to meet teachers' needs? Keep in mind that there are companies that can help you with designing customized professional learning opportunities so you don't have to go it alone. Remember that you are a mentor to your teachers with regard to instruction; you are charged with helping teachers think differently about their instruction.

### *Better Math Education*

Better math education is about more than just memorizing equations—or in a broader sense, more than about a given type of pedagogy. Better math education requires us to help students embrace and appreciate math. It's about helping students develop a positive perception of math and refusing to accept that math isn't for them, or that they will never be good at math. To change math outcomes in our society, we must first change math perceptions. No textbook or resource will ever help students if they do not first open their minds and hearts to learning, and that starts with redefining what it means to learn math.

# AFTERWORD

Shortly before this book was published, I (Brian) was asked, "What does learning look and feel like to you?" After giving it some thought, I provided my response and very quickly realized that my response was actually a process.

- We learn by living; we learn by doing.
- Learning is a series of trials and errors; we are not perfect.
- Learning is messy, because messes can only be created through activity.
- When we are active, we are engaged.
- When we are engaged, our minds are busy constructing and organizing information.
- Constructing and organizing information is learning!

To my fellow administrators, teachers, parents, and students, the next time you see a mess, try your best to see it for what it really is: ***a piece of the learning process.***

You have one life. Help others. Whether you are an administrator, teacher, parent, or student, look for opportunities to make the world a better place. People very rarely look back on their lives and say, "I wish I would have done less." Doing more might not be easy, but it's worth it.

If you found our book to be helpful, please share it with a friend or colleague.

Thank you!<br>
Brian and Kris

# References and Resources

Abram, Karen et al. "Post-Traumatic Stress Disorder and Trauma in Youth Juvenile Detention." *Archives of General Psychiatry,* 61 (2004).

*The Art and Science of Teaching: A Comprehensive Framework for Effective Instruction* (Professional Development) 1st Edition. Association for Supervision & Curriculum Development (2007).

Baer, Drake. "How Your Mindset Determines Your Success, Well-being, and Love Life." *Business Insider* (August 20, 2014)

Boaler, Jo and Dweck, Carol. *Mathematical Mindsets: Unleashing Students' Potential through Creative Math, Inspiring Messages and Innovative Teaching,* 1st ed. Jossey-Bass (2015).

Brown, Louise. Elementary Teachers' Weak Math Skills Spark Mandatory Crash Courses. thestar.com. (May 13, 2016)

Caldji, Christian et al. "Maternal Care During Infancy Regulates the Development of Neural Systems Mediating the Expression of Fearfulness in the Rat." *Proceedings of the National Academy of Sciences 95*, no. 9 (April 28, 1998).

Caldji, Christian, Josie Diorio, and Michael Meaney. "Variations in Maternal Care in Infancy Regulate the Development of Stress Reactivity" *Biological Psychiatry* 48, no. 12. (December 15, 2000).

Castelvcecchi, David. "'We Hate Math,' Say 4 in 10—a Majority of Americans." *Scientific American* (October 15, 2011).

Ellis, Arthur K. *Exemplars of Curriculum Theory*. Larchment, NY: Eye on Education. (2004).

Evans, Gary and Schamberg, Michelle. "Childhood Poverty, Chronic Stress, and Adult Working Memory" *Proceedings of the National Academy of Sciences* 106, no. 16. (2009).

Fogg, B. J., *A Behavior Model for Persuasive Design*, white paper, Persuasive Technology Lab, Stanford University. www.mebook.se/images/page_file/38/Fogg%20Behavior%20Model.pdf. (2009) Also see www.BehaviorModel.org.

Frontier, Tony. "Making Teachers Better, Not Bitter: Balancing Evaluation, Supervision, and Reflection for Professional Growth." Association for Supervision & Curriculum Development (2016).

Gardner, H. *Intelligence Reframed. Multiple Intelligences for the 21st Century*. New York: Basic Books (1999).

Harvard Education Center for Policy and Research. *Mathematical Quality of Instruction* (MQI) https://cepr.harvard.edu/mqi.

Joyce, B., Weil, M., Calhoun, E. *Models of Teaching* (8th ed.), New Jersey: Pearson Education Inc. (2009).

Liu, Doug et al. "Maternal Care, Hippocampal Glucocorticoid Receptors, and Hypothalamic-Pituitary-Adrenal Responses to Stress." *Science* 277, no. 5332 (September 12, 1997).

Mathnasium of Littleton. *How Elementary Teachers are Set Up For Failure in Math*. (July 2, 2016).

# References & Resources

McEwen, Bruce. "Protection and Damage from Acute and Chronic Stress." *Annals from the New York Academy of Sciences* 1032. (2004).

Ostashevsky, Luba. "Elementary School Teachers Struggle with Common Core Math Standards." *The Hechinger Report* (June 15, 2016).

Ornstein, Allen, Hunkins, Francis. *Curriculum: Foundation, Principles, and Issues* (5th ed.). New Jersey: Pearson Education Inc. (2009).

Peters, Brian A. *The METUS Principle: Recognizing, Understanding, and Managing Fear.* HenschelHaus Publishing (2015).

Peterson, Paul. *Are U.S. Students Ready to Compete?* EducationNext Fall 2011 / Vol. 11, No. 4.

Reardon, Christopher. *Research on School Reforms.* Quasar: A Brighter Future in Mathematics. http://creardon.com/archives/FFR/FFR16.html. (1996).

Robelen, Erik. "U.S. Math, Reading Proficiency Falls Short in Global Analysis." *EdWeek.* August 19, 2011. http://blogs.edweek.org/edweek/curriculum/2011/08/us_math_reading_achievement_fa.html.

Stein, M., Smith, M., Henningsen, M., & Silver, E. *Implementing standards-based mathematics instruction: A casebook for professional development.* New York: Teacher College. (2000).

Stiger, James. *The Teaching Gap: Best Ideas from the World's Teachers for Improving Education in the Classroom.* Free Press; Reissue Edition (June 16, 2009).

Tough, Paul. *How Children Succeed: Grit, Curiosity, and the Hidden Power of Character.* Mariner Books (2012).

Trei, Lisa. "New Study Yields Instructive Results on how Mindset Affects Learning." *Stanford Report.* (February 7, 2007).

Wingert, Pat. "When Teachers Need Help in Math." *The Atlantic* (October 2, 2014).

Wood, S.E., Wood, E.G., Boyd, D.A. *The World of Psychology* (3rd ed.) Allyn and Bacon Publishing (2007).

# ADDENDUM A

### PRACTITIONER'S SUMMARY

Just as we know that student change doesn't happen overnight, it is unrealistic for us as practitioners to change our practices 100 percent in one year.

As you embark on your journey to help students love math—or at least not fear math—take some small steps. As teachers, we need to make sure that our classrooms are engaging, pleasurable, and safe environments that help develop meaning and a deep conceptual understanding for students. To do this, we need to keep the *Eight Standards for Mathematical Practice* in the forefront.

Regardless of your political views, *Common Core Standards for Mathematical Practice* (SMPs) are what is best for learners. These standards are written in a common language for all educators and are not just *what* (content standards) should be taught, but *how* (practice standards).

These math practices are based on solid research from two main resources: the National Research Council's Five Strands of Mathematical Proficiency (*Adding It Up—Helping Children Learn Mathematics*) and the National Council Teachers of Mathematics (NCTM) Process Standards.

### NRC's 5 Strands of Mathematical Proficiency

- Adaptive Reasoning
- Strategic Competence
- Conceptual Understanding
- Procedural Fluency
- Productive Disposition

### NCTM's Process Standards:

- Problem Solving
- Reasoning and Proof
- Communication
- Representation
- Connections

# Addendum B

## Eight Standards for Mathematical Practices (SMP):

In accordance with these standards, mathematically proficient students will:

1. Make sense of problems and persevere in solving them.
2. Reason abstractly and quantitatively.
3. Construct viable arguments and critique the reasoning of others.
4. Model with mathematics.
5. Use appropriate tools strategically.
6. Attend to precision.
7. Look for and make use of structure.
8. Look for and express regularity in repeated reasoning.

For students to be fluent in all of these mathematical practices, they will need to have time to use them regularly. Our state assessments are changing to incorporate more rigorous tasks and higher-order thinking. As teachers, we must allow students to practice several of these eight standards every day so they can become proficient. A higher-order thinking task will usually provide students with opportunities to use several of the standards at one time.

**1. Make sense of problems and persevere in solving them** can be practiced daily in classrooms. In the beginning, students will need to be trained on how to accomplish this. Here are some student behaviors to keep in mind when looking at this practice:

- Students will need to explain their reasoning.
- Students should be able to connect mathematics and provide multiple representations for the same scenarios.
- Students should be asking themselves and their peers, "Does this make sense? How is my method similar or different?"
- At first, teachers will need to ask students to explain their reasoning. As students start understanding the expectation, they will automatically provide their reasoning and explanations.
- Elementary and some middle-school students will need to use concrete objects or pictures to support conceptualizing and solving the task.
- This standard promotes curiosity to make conjectures and formulate a plan without just focusing on getting an answer.

> *Reducing anxiety is a great way to enhance and foster the willingness to persevere.*

## Addendum B

***Suggested Teacher Moves:***

- Allow students "think" time first before they share with their groups in order to formulate their own reasoning.

- Allow students ample time to collaborate. Talking will be necessary! You will need to listen, ask clarifying or prompting questions, and stay out of the way once the students get the hang of persevering.

- Ask "why" and "how" questions.

- Provide students with time and encouragement, especially if students have a fear of math.

Each year that students are encouraged to ***make sense of problems and persevere in solving them***, the easier the mathematics collaboration at the beginning of a school year will be. It is easier to get students engaged earlier, and start developing good learning behaviors earlier, than it is to "fix" undesirable learning behaviors later in life. The foundation for good learning is not just content knowledge, which state departments of education have pushed in various states in recent years, but rather in establishing an environment in which students develop a learner's mindset as well as practice corresponding behaviors.

> *What's easier, engaging students early, often, and ongoing...*
> *or re-engaging students after they have developed poor habits*
> *and a negative sense of self as it pertains to being*
> *"a math student"?*

**2. Reasoning abstractly and quantitatively** can be a lot of fun for students. For students to become proficient with this practice, they will explore real world scenarios. As a teacher, select tasks that interest your students. The first part of this practice—the ability to reason abstractly (decontextualize)—means that students can read and understand what the scenario is asking and represent it symbolically.

We find that starting most days with a task that involves a scenario really helps students practice this standard. After several months, they are no longer afraid to try. It's also a great way to not just focus on the answer, which will require perseverance.

The second part of this standard—to reason quantitatively—requires students to contextualize. Students who love writing enjoy contextualizing once they get the hang of it. A teacher could write an equation on the board, for example, $24.95s + 49.95 = T$ and ask the students to write a story that represents this equation. There is no one right answer.

Students are able to be creative and as they share their "context" with the group, the class can decide whether or not their story makes sense. Students might write about their cell phone plans, selling t-shirts, renting cars, purchasing food, and other things that interest them. This practice is focused on making meaning of quantities, not just getting an answer.

## Addendum B

***3. Constructing viable arguments and critiquing the reasoning of others*** requires students to argue mathematics, which is something that should be happening each and every day in your math classrooms. The *Teaching Channel* has an instructor who models this standard by starting her classes with, "My Favorite No." (teachingchannel.org) Watch it.

In elementary grades, students need to construct arguments using objects, diagrams, drawings, and discussion. As students mature, they will generalize more and determine domains in which their answers will be correct or incorrect. Teachers have to provide a safe environment in the classroom for students to productively struggle and converse. This practice is not about getting a correct answer; its focus is to have students reason and justify their methods and the methods of others.

This mathematics practice standard will also bring a lot of laughter and fun if you allow it to, because students have the most creative and fun time when creating arguments. They will solve scenarios based on their likes and dislikes. For example, you could give the class three different choices for going on vacation—driving, flying, or taking the train. Give them the costs, distance, and times, and ask students to decide which one is best. Let the students solve for each mathematically and share their ideas.

When Kris once did this exercise, one student said she would fly because her parents were paying for the vacation and they didn't care about the cost. Another student said he had just gotten his driver's license, so he wanted to drive. Another said flying was not an option because her dad would never fly. Students' anecdotes will be entertaining, and making math personal will have meaning for them.

4. When *modeling with mathematics*, teachers tend to have some confusion. This practice standard goes hand in hand with decontextualizing in the previous standard but takes it one step further—expecting students to provide a correct solution. Here are some of the expectations for students:

- This standard requires students to analyze and synthesize to the maximum degree.
- The students will have to determine what the task or problem is asking them and apply their previous mathematical understandings in order to work towards a solution.
- Students will determine whether their models are working and if they need to make improvements.
- Students will constantly check and re-check to determine if their methods and answers make sense.
- When students master this standard, they are working at the highest level of cognitive thinking.

When students are modeling with mathematics, teachers will need to allow time. Tasks where students practice this standard will be rigorous and robust in nature. Projects can

# Addendum B

be one way of allowing students to practice modeling with mathematics, and many teachers create grading rubrics for projects. Just be careful that your rubric is generic enough not to lower cognitive thinking or suggest models.

This is the time for teachers to truly become that "guide on the side" and be facilitators of learning. Allow students time to create their own models and form their own thoughts before collaborating with their peers.

5. When students *use appropriate tools strategically*, they select tools they can use to solve mathematical tasks. As a teacher, you should introduce or continue to redistribute tools students have been using in previous grade levels. If you aren't certain what those tools might be, take the time to figure it out. Ask the teachers two grade levels below and above what manipulatives they are using with students.

Manipulatives might be one kind of tool, but don't forget about other tools like software, calculators, rulers, protractors, arm lengths, foot lengths, string, and so on. The idea for students to become proficient in this standard is that they can pick the most efficient tools to solve the task. Students will use these tools to create or support the creation of their mathematical models in Standard 4.

Allow students to use tools so that they can figure out which tools make the most sense when solving problems. Teachers need to guide students to appropriate tools at times so they do not to veer off course too much.

Try Kris's number line task by taping register tape on the desks and providing students with index cards with 22 different real numbers. Have students determine the order and placement for each number on the register tape without changing its size, using a ruler, or taking the tape off the desks. Don't provide them with any tools; see what they find around them to measure.

Like the students in Kris's class, they might use binders or notebooks to determine the total length of the tape, then then divide it into equal parts. Or use the arm of a student or even a shoe.

6. **Attending to precision** can only be achieved if teachers themselves are modeling this standard on a daily basis. Attending to precision starts with precise communication.

We are always told that mathematics is a language, and this standard helps support that statement. Elementary teachers need to use the correct mathematical terms so that students aren't confused as they move into middle and high school. Teachers need to set the expectation that numbers aren't "naked"—they should always have units. Students will meet whatever expectation and example you model for them. Encourage students to use proper terms, use clear definitions, describe their symbols, and specify the units written and orally.

When students attend to precision, they take pride in their work and hold each other accountable. After some

# ADDENDUM B

reminding about units and precise explanations, they will correct each other if one forgets the units in their answers.

This mathematical practice standard must be applied daily, or it will be tough for students to become proficient. Be careful not to lose your safe environment and go overboard with correcting students. If math anxiety is prominent, consider thanking students for using precise language and encouraging rephrasing if the definitions aren't quite up to par.

**7. *Look for and make use of structure*** is closely related to mathematical practice standard 8, ***look for and express regularity in repeated reasoning.*** When students are proficient with SMP 7, they will notice a pattern or structure. Most state content standards include patterning, when in fact, finding patterns is one method of solving mathematical tasks.

Students start looking for structure and patterns at very young ages. They will usually ask a lot of questions and will want to know if their hypotheses are true. Allow students to explore their conjectures to determine correctness. This encourages students to continue to look for structure in mathematics. When they become proficient, they will make use of this structure to determine limits on answers.

**8.** When students become proficient at mathematical practice standard 8, ***look for and express regularity in repeated reasoning***, they will notice what is happening in mathematical practice standard 7 and look for shortcuts. They will test out their repeated reasoning to make sure that it always works, and then they will use the shortcut they created. Students need to continually attend to precision and check to make sure their reasoning makes sense.

# Addendum C

## Business Card Activity

*Purpose: To promote non-mathematical discourse*

The Business Card Activity provides students with an opportunity to get to know something about one another. Students are typically more willing to tell one person about themselves as opposed to sharing with a larger group. It opens up the opportunity for non-academic communication, an important step before academic communication.

Hand out colored index cards to the students, and have them pair up with someone who has the same color card. Make sure to only hand out even amounts of each color so that everyone has one partner. If you have an odd number of students, then have one group with three members, and have them interview each other in turn—combining all three steps of the process.

Tell the students they are to interview their partner and make a business card for them. Show the model business card and point out each part.

Note that the "symbol" is what the interviewee describes to the interviewer as something that represents who they are (their "identity"). For example, a person who is musical may describe a music note as their symbol. The interviewer draws the symbol as the interviewee describes it. The symbol is similar to a logo that businesses or individuals often put on business cards. Be sure to mention that symbols need to be school-appropriate.

Have Student 1 interview Student 2. Give students instructions for the first two steps. You should time this activity, usually 5 to 7 minutes, and then have them switch roles for the interviews.

**Step 1:** Student 1 interviews and makes a business card for Student 2.

**Step 2:** Student 2 interviews and makes a business card for Person 1.

| If you were not here, where would you most like to be? | Symbol | How many brothers or sisters do you have? |
|---|---|---|
| | **First and Last Name** | |
| What is your favorite subject? | | What do you like to do in your spare time? |

# Addendum C

Monitor the groups, and give a one-minute signal when you see the first partner pairs finishing.

Give instructions for the next step. Have students pair up with another pair to form a group of four. Ask whoever is the youngest in the group to start, or some other random method. Tell them they will have about 8 minutes (2 minutes per student) total to do this next step.

**Step 3:** Each student in the group introduces the person s/he interviewed to the rest of their small group using the business card answers as a basis for introduction.

Give a signal at the end of each two-minute segment to move on to the next introduction to help the group pace their process.

**Extension:** Have students write their email addresses and phone numbers on the backs of their business cards, and give them to their new "homework partners." If they should ever need help or have questions on homework, they should contact their homework partners.

# About the Authors

Brian A. Peters is an award-winning author and professor at the Milwaukee School of Engineering in the Rader School of Business. As an author, speaker, and consultant, his work centers on improvement, systems, and outcomes. He received his B.A. from the University of Wisconsin–Whitewater, where he studied psychology; his M.B.A. from the University of Colorado at Colorado Springs; his M.S.A. from Central Michigan University; and his M.Ed. with an emphasis on Curriculum and Instruction from California Coast University. Brian currently lives in Wisconsin with his wife, Missy, his daughters, Kensington, McKinley, and Leilani, and their family dog, Sadie.

Kristine E. Hobaugh is a director of professional development and manages a team of professional developers to support school partnerships. She taught high school mathematics in a suburb of Pittsburgh for 10 years and served as adjunct professor at the University of Pittsburgh at Greensburg for several years.

Passionate about the teaching and learning of students, teachers, and administrators, she delivers workshops and speaks across the USA.

Kris earned her B.S. in Secondary Education Mathematics with a minor in Computer Science at the University of Pittsburgh and received both her M.Ed. in Mathematics and Computer Science and K-12 Principal Certification from California University of Pennsylvania. Kris currently lives in Ligonier, Pennsylvania with her husband Don and son Nathanial, who is the star of some of her stories.

CPSIA information can be obtained
at www.ICGtesting.com
Printed in the USA
FSHW020918150519
58136FS